Maria Aparecida de Sá Xavier

Ticumbi and the Art of Healing

Maria Aparecida de Sá Xavier

Ticumbi and the Art of Healing

Expressions of spatiality in the village of Itaúnas

ScienciaScripts

Imprint
Any brand names and product names mentioned in this book are subject to trademark, brand or patent protection and are trademarks or registered trademarks of their respective holders. The use of brand names, product names, common names, trade names, product descriptions etc. even without a particular marking in this work is in no way to be construed to mean that such names may be regarded as unrestricted in respect of trademark and brand protection legislation and could thus be used by anyone.

Cover image: www.ingimage.com

This book is a translation from the original published under ISBN 978-3-330-77074-4.

Publisher:
Sciencia Scripts
is a trademark of
Dodo Books Indian Ocean Ltd. and OmniScriptum S.R.L publishing group

120 High Road, East Finchley, London, N2 9ED, United Kingdom
Str. Armeneasca 28/1, office 1, Chisinau MD-2012, Republic of Moldova, Europe
Managing Directors: Ieva Konstantinova, Victoria Ursu
info@omniscriptum.com

Printed at: see last page
ISBN: 978-620-8-63365-3

SUMMARY

1 Introduction

This book presents an unprecedented investigation into the practices and knowledge of the art of healing in the rural community of Vila de Itaùnas, understood here as an expression of spatiality, a *modus vivendi*, carried out between 2005 and 2008. We understand that the geographical space we are looking at and its respective spatiality are social entities. In this sense, space is not just a support, substrate or receptacle for human actions, not to be confused with the physical base, but a produced space, as Moreira (2007, p.64) warns us, and spatiality is understood as the concrete empirical state of organization of the form of spatial existence of entities (MOREIRA, 2008, p.55). The purpose of this work is therefore to understand how spatiality is constructed by the social actors in Vila de Itaùnas, why, how and where it appears within the scope of the investigation. However, we do not intend to surrender to the "seduction of the village", as ethnographers say, since the "village" is a complex 'place' and within this complexity, it is impossible to account for all the nuances. However, with a proposal for ethnogeographic work, with a view to capturing the socio-cultural aspects of the space, we have sought a synthesis as far as possible within the theoretical limits of this work.

We start from the premise that man's being in the world is spatial, dwelling, *and language* is the dwelling place of being, in the thought of the philosopher Heidegger (1995, 2002), and Gadamer (1996) perfected the master saying "being that can be understood is language"[1] . This way of spatializing, *inhabiting* embodies an ethos in a territory, making it a place like one's own, since being-ai - *dasein* - being, in and through *language*, expresses a *modus vivendis*. In another way, I can say that the *way of living* can be read and perceived as a conniving Landscape, understanding landscapes as texts that can be read. From this perspective, landscape is perceived as *connivance,* in Gilles Sautter's (1978, apud CLAVAL, 2004, p.49) reference to a landscape that goes beyond an objective reality, and presents itself as a mode, a way in which this reality speaks to the senses of those who discover it, the way in which it comes into harmony with their states of mind, or goes against their moods. In this sense, it is necessary to explore the *conniving landscape* (BONNEMAISON, 2002, CLAVAL, 2004) in order to perceive the crossed wires, the reciprocal exchanges, an appearance and a representation that do not appear at first glance, because they are invisible. These forms of fixity and flow, meanings and signifiers are embodied as geosymbols, as references of a space/time, presented and represented in space by and through the everyday life of the social group, in this case the community in Vila de Itaûnas, a community of belonging[2]

By understanding *dasein* as *being-there, being* in its daily life, we have chosen to work with a variety of concepts such as spatiality, territoriality, place and identity, as we understand that these express *moments of being in the world,* within the cutout we wish to make guided by the gaze.

Everyday life presents itself as a reality that can be interpreted by humans and subjectively endowed with meaning for these humans in a state of community insofar as it forms a coherent whole, according to Berger and Luckmann (1996). The object of study - spatiality and its expressions in the art of healing - is to be found

1 A language beyond linguistics. It's no coincidence that Michel Foucault (2004) looked at discursive forms, enunciations, to understand the power games that underlie *events*.

2 In the Weberian sense of belonging to a place. We belong to something that belongs to us.
M. Santos in a lecture in the auditorium of the Institute of Geosciences, UFF, which bears his name, summer 2000.

in the reality of everyday life, which originates in the thoughts and actions of ordinary human beings and is affirmed by them as real, and so we will *interpret* everyday life and the landscapes in the village of Itaúnas. This reality is also intersubjective, since the world in which the social actor in Vila de Itaûnas participates is in confluence, like a social network, with other actors where it exists continuously in interaction and communication with others, and with the world.

According to Berger and Luckmann (op.cit., p. 43), religious experience is rich in producing theatrical transitions (where there are transitions between realities - the staged world and the world of everyday life), to the extent that art and religion are endemic producers of fields of meaning. Following this path of thought, we present the religious institution of Ticumbi - a hierarchical, religious association, but not only that, which presents its religious identity, its faith, as a gathering of congos who manifest themselves in commemorations, rehearsals, rituals and celebrations of St. Benedict and St. Sebastian, the latter being the patron saint of the town of Itaûnas. Ticumbi is a category invented by Luiz Guilherme Santos Neves, a historian and folklorist from Espírito Santo, to differentiate the congo from the northern region of Espirito Santo from other congos. According to Medeiros (2008), the members of the congo and the people who remain from quilombos in the north of Espírito Santo only call it Baile de Congo. Ticumbi[3] is considered one of the most important manifestations of *folklore* in Espírito Santo, and according to Medeiros (ibid.) the term may have been used as a corruption of Cucumbi. Cucumbi exists in the Northeast and Cacumbi in Rio de Janeiro; in this case, mestre Guilherme changed the original name from Baile de Congo to Ticumbi, although the more traditional ones continue to refer to the festival as Baile de Congo[4] . However, as the interviewees say Ticumbi, we will use this category to refer to the celebration, the procession. We'll go into more detail in another chapter.

In order to shed light on the conceptual basis, we need to understand, even briefly, what we call a social institution. A social institution (BERGER AND LUCKMANN, op.cit.) originates from habit, since any action that is repeated frequently becomes molded into a pattern that can be reproduced with economy of effort. Habit also implies that the action can be performed in the future, and in the same way. These processes of habit formation precede any institutionalization. Institutionalization occurs whenever there is reciprocal typification, and the typifications exercised in this reciprocity give the typical character not only to the actions, but also to the actors in the institutions, since these typifications are "shared" with others and are accessible to all members of the particular social group in question - in this case the Ticumbi of Vila de Itaûnas. We can infer that if institutions have a historicity, they also have a spatiality and function as social control - they operate a territoriality, they inaugurate a territory. For Berger and Lukmann (ibid) it is very important to understand the historical conditions in which a given institution emerged. To summarize, the authors tell us that "the institutional world is objectified human activity, and this in each particular institution", and that the relationship

3 Traditionally, this folguedo or dance is made up of a certain number of members and is performed in commemoration of St. Benedict. It is staged with the following characters: the Congo king, the Bamba king, their respective secretaries and the corps de ballet of each nation, representing the warriors. The plot of the play revolves around the dispute between the two kings who want to hold the feast of St. Benedict separately. The challenges are daring and provocative, declaimed by the secretaries, but in keeping with tradition, the issues dealt with in the verses are always up to date.
4 Zaluar (1983) shows that the celebrations to commemorate St. Benedict always end with a dance, like forró. In the village of Itaunas, this is no different.

between man, the producer, and his social world, his product, remains a dialectical relationship. In other words, man and his social world reciprocally act on each other dialectically. The institutional world also requires legitimation, that is, the ways in which it can be explained and justified. For this, there are 'legitimizing formulas', since they bring the institutional order to new generations, and so form conviction, through which they exercise a certain authority over the individual. The whole process is shared with others, along with its meanings, which are integrated into the common biography of the individuals in this social state.

Every society has a *corpus*[5] of knowledge, of arts and crafts, which provide a composition of understanding within which the subject is provided with instruments to *move in the world*[6] , and what is not yet known will become known in the future. In this sense, Foucault (2004, p.205) tells us that *there is knowledge that is independent of the sciences; but there is no knowledge without a defined discursive practice, and every discursive practice can be defined by the knowledge that it forms* and informs:

"A knowledge is what we can talk about in a discursive practice that is specified as follows: the domain constituted by the different objects that will or will not acquire a scientific status; (...) a knowledge is also the space in which the subject can take up a position to talk about the objects he deals with in his discourses; (...) a knowledge is also the field of coordination and subordination of the statements in which the concepts appear, are defined, applied and transformed; (...) finally, a knowledge is defined as the field of coordination and subordination of the statements in which the concepts appear, are defined, applied and transformed....] knowledge is also the field of coordination and subordination of statements in which concepts appear, are defined, applied and transformed; (...) finally, knowledge is defined by the possibilities of use and appropriation offered by discourses." (FOUCAULT, 2004, p.204)

In these terms, the Ticumbi is a religious and political association in the village of Itaûnas, in the district of Conceiçâo da Barra. It is a true institution of popular Catholicism with its knowledge and practices: its specialists, the *congo masters*, the *experts in the art of healing* such as the *benzedeiras, the midwives and the rezadores*.

Although we are discussing the Ticumbi, it won't be our aim at the moment to describe this rich religious and cultural institution in detail, but rather to show in the interplay of social positions, how and where the specialists in the knowledge of the art of healing are inserted, already as a sub-institution, in order to give an understanding of these levels of social organization. If we don't do this, we run the risk of losing touch with this real invisible *social network*, which is half-embedded in the landscape and carries signs. These networks woven by the social actors in Vila de Itaùnas organize the space of everyday life, and the extra-everyday life of the festivals, through their solidarity and affectivity through faith. A sacred territory is founded in everyday space, a hierophany, since this fabric permeates all relationships, organizing calendars, processions, festivals, and marks a time with recurring social practices - traditions, presented in subsequent chapters.

To clarify, we call *specialists* those who have a knowledge/doing, a technique, an art, a knowledge about the origins, the singing, the choreography, the "secret", the magic, the faith, the magic words, and the rites of

---

5 A set of enunciations, forms and discursive practices that exceed on all sides (FOUCAULT, 2004, p.200-2001)

6 Building their homes, foraging for food, taking care of their health, providing themselves and their families with the goods they need to survive, and a system of beliefs and symbols that sustains their daily life of values, Bourdieu's symbolic capital (1996).

passage[7] that will be passed on to the next generation through the process of oral tradition (preferably, but not only as we shall see). Berger and Luckmann (op. cit.) inform us that all transmission requires some kind of social apparatus in order to be carried out and that is why this body of specialists is of great importance, since it becomes a reference for the knowledge accumulated in the process of transmission.

In an institution, everyone plays social roles, and it is in this way that the individual participates in the social world; each role represents the institutional order. This implies that there is a social distribution of knowledge, and in this way we can talk about the social prestige of specialists, agreeing with Michel Foucault when he says that all knowledge contemplates power, since all 'knowledge' has its genesis in power relations, within a process and in various fields of knowledge. Power in this case is a set of open relationships, more or less coordinated (poorly coordinated), and appears as an operation of political technologies. We are not talking here about party politics (even if power falls into the State's mesh, is appropriated by it, and uses it, de-characterizing its essential core), but about political instruments, which, through and in the relationships, discourses and forces engendered by the agents, give rise to the "event"

Foucault (1995, p. 242, 243) explains that power

"(...) it is a set of actions on possible actions; it operates on the field of possibility where the behavior of the active subjects is inscribed; it incites, induces, diverts, facilitates or makes difficult, amplifies or limits, makes more or less probable; at the limit, it absolutely coerces or prevents, but it is always a way of acting on one or several active subjects, and the extent to which they act or are susceptible of acting. An action on actions. (...) The exercise of power consists of "driving behavior" and ordering probability. Power, in essence, is less about the confrontation between two adversaries, or the attachment of one to the other, than it is about "government". (...) When we define the exercise of power as a mode of action over the actions of others, when we characterize it as the "government" of men by one another - in the broadest sense of the word - we include an important element: freedom. Power is only exercised between "free subjects", insofar as they are "free" - by which we mean individual or collective subjects who have before them a field of possibility in which different behaviors, different reactions and different modes of behavior can take place. (...) There is therefore no confrontation between power and freedom, in a relationship of exclusion (where power is exercised, freedom disappears); but a much more complex game: in this game, freedom will appear as a condition for the existence of power (...)."

Here we come to the point in this introduction where we understand that the Vila de Itaûnas community constructs a spatiality through its daily life, affirming a territoriality (we'll conceptualize this later), an identity, a *modus vivendi*. This way of being in the world is supported by a religious institution of popular Catholicism, unrecognized by the clergy, where the sacred and the profane coexist and intermingle at various times. This social institution operates through its agents who play their roles. These roles are legitimized through political, signifying and power operations.

It will not be possible to forget (using the force of repetition) that here in this work we are not trying to understand the Ticumbi Institution, nor to describe it in detail and its syncretic diversity, but rather to understand through the actions of the agents, in the dynamic process, how these actors articulate themselves,

7 Rites of passage according to Van Gennep (Turner, 1974, p. 116) are "rites that accompany every change of place, state, social position and age". Rites of passage are characterized by three phases: separation, margin (limen, or threshold in Latin) and aggregation.

building a spatiality, a differentiated territoriality, founding a community of belonging. It is in this sense that *the 'healing know-how' of* this community will be understood here as *expressions of this spatiality, this modus vivendi.* This knowledge is contained within a larger *body of knowledge,* which gives meaning to and explains the world through faith. As well as understanding that this can only be achieved through power games, and power being a way in which some people act on others in relationships and through social actions, given that the *art of healing* is a sub-institution, certainly religious, but also and no less political, of so-called traditional or popular Catholicism and very well described by Zaluar (1983) and Steil (2001, p. 9-40), as we will see in a later chapter.

1.2 In search of a path.

The aim of the method adopted was to map[8] the symbolic representations and aspects of doing things with knowledge (*corpus*) of the health/disease/healing process in Vila de Itaùnas, C. da Barra, ES, as expressions of a social spatiality, a *modus vivendi.* To this end, we constructed levels of understanding of the object in its subjectivity and apprehended reality (i) a theoretical survey of conceptual aspects in the form of an essay; (ii) the creation of a mental map and an interpretative map. Through the maps we sought to understand how the community constructs its space/place and dwelling through its social networks; (iii) bringing the two realities closer together, academic and local heritage knowledge in the exchange of knowledge (also constituting a network); (iv) understanding the social spatiality expressed in the doings with health/illness/healing knowledge, as well as its sacred/profane space within its daily life or mundanity.

The ethnographic method was used in order to capture the subtle data of understanding the spatiality of the society in question. The interpretivist ethnographic methodology[9] , "participant observation[10]and the "estrangement"[11] of behavior and representations (speech, stories, art, magical/religious rituals, songs, dances and games) completed the method. Twenty open-ended interviews (recorded) were carried out with local actors, using a memorized script. However, there were many conversations that were not recorded, but which were noted in the field diary and lent themselves as input for interpreting the field. Photographs and short films were taken as a representation of a space/time that helped to reconstruct the data included in the interpretative map, as well as to understand the social network. Among the twenty interviewees were men and women aged between 16 and 85, and the specialists in healing (benzedeiras, midwives and rezador) indicated by the local actors were the privileged informants. The PEI stakeholders interviewed were: administrator André L. Tebaldi and Luciana Franco Verissimo. The local actors who work in the PEI were part of the investigation, but were not interviewed formally, but informally, in a chat .[12]

---

8 This cartography goes beyond a geographical mapping of a static territory. In fact, it is a drawing that accompanies the changes and meanings produced by certain life situations in the territories lived and thus constituted as "place" (SEEMANN, 2006; HOLZER & HOLZER, 2006). Cartography, in this sense, seeks to capture the universes traced by ways of life, by desire, by the representations of this *being in the world* and how they construct their signifiers and meanings. Beyond structuralism and functionalism, we have focused on the signifiers of Geertz (1989, 2002).

9"Examining the world through those who speak of it - how it is portrayed, lived, demarcated, represented - and not through what it intrinsically is," according to Geertz (1997, p.ll).

10 On the ethnographic method, see Cardoso de Oliveira (2000, p. 17-35).

11 See Lèvi-Strauss (1993, p. 278).

12 This was the best way to do it, since in an informal chat there are jokes where the social actor feels more relaxed,

We used the ideas of Seemann (2006), Kozel (2006), Wherter Holzer & Selma Holzer (2006) and Carvalho (2006) to create the mental map. These authors show that the experience of lived space can be represented through a so-called mental map. Cartography, according to Holzer & Holzer (2006, p. 201) is an act of intersubjective communication, and also a way of placing oneself in the world, the art or science of representing it, in order to orient oneself, bring the there to the here, make space familiar and turn it into a place. A map is a representation of the territory, but not a passive representation. It expresses, presents a conception of the world, an organization of the lived world, of lived space, but also of imaginary space. Concrete places and imagined places. A map represents power and knowledge of lived territory, a worldview, and essentially a cultural representation (SEEMANN, 2008, NEPEC Symposium table presentation).

Along these lines and in order to understand this social spatiality, a mental map was constructed by 16-year-old Lucas André Maia dos Santos, a third year student at the Benônio Falcao Gouveia School[13] , with the collaboration of this author. The choice of student was not random, as he had been recommended by local social actors in the village as someone who knew how to build a map, and had already done so in the school's geography class.

The map shows the representation of lived space made by this social actor, as a representative of one of the oldest families in the village and who actively participates in all cultural events, as well as activities in the Itaùnas State Park. The first map was drawn on cardboard using a black pencil. This map was then digitized with the help of geographer Renato Gomes S. Barcellos (UFF) and worked on in the Coreldraw program with the help of geographer Tatiana de Sà F. Ferreira (UFRJ).

The mental map shown in figure 1 sought, at first, to represent the social space of the village, the location of the homes of the healing specialists: benzedeiras, rezadores and midwives and the main socio-spatial references, including the geosymbols in the village. The following were identified as social references of a time/space: Health Post, Church of Sao Sebastiao and its annex, Church of Sao Benedito (unofficial), Square with its Pequi Vinagreiro trunk in the center, Police Station, Pousada Dunas (the first pousada in the village), the Benônio Falcao Gouveia School, Cemetery, Football Field, Casa do Cidadao (where the post office is), Fishermen's Association house, Forró Buraco do Tatu, Bar Forró (the oldest forró bar), Sports Court, Pharmacy, D. Teresa's Restaurant, Pedrolina's Restaurant (Pedrolina). Teresa's Restaurant, Pedrolina's Restaurant (Teresa's sister).

avoiding that observer x observed position.

13 Known by all the community members as the "Benonio School".

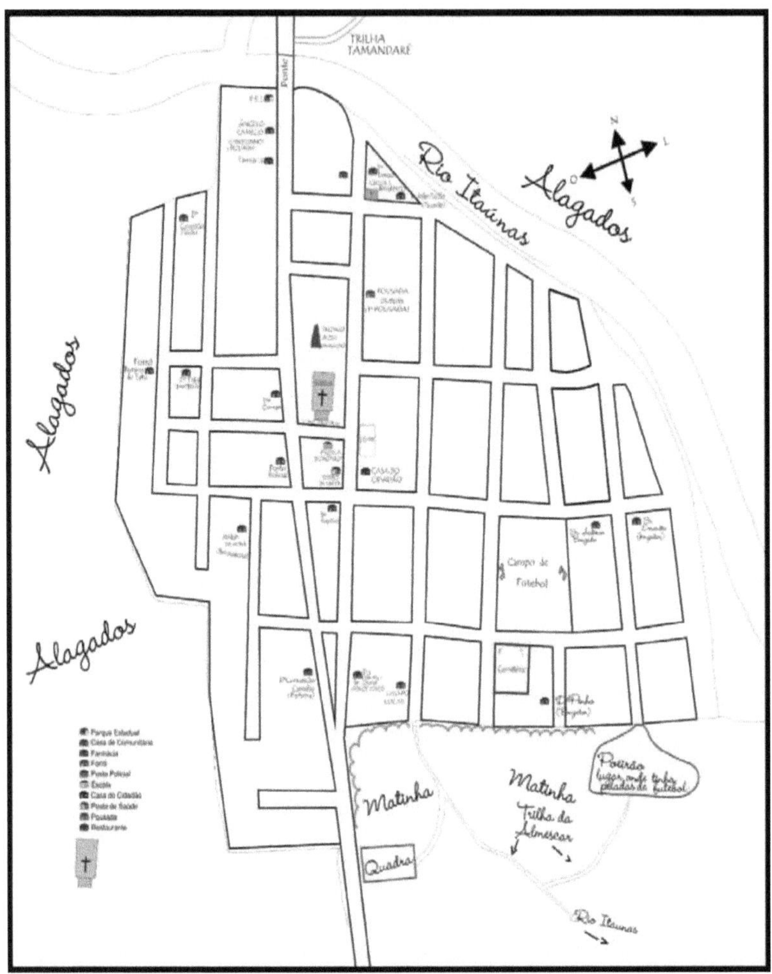

Figure: mental map, version 1, made by Lucas André Maia dos Santos (16 years old), representing the social space of Vila de Itaûnas, in 2007. Map digitized by the geographer Renato Gomes S. Barcellos-UFF, being worked on in Coreldraw by the geographer Tatiana de Sà F. Ferreira-UFRJ, with idealization and method by Xavier (2008).

In a second step, another map was built using the same references as the mental map, but augmented by references from the ethnogeographic work. The experience of the fieldwork, through participant observation and informal conversations, were relevant to the construction of this map, called here the interpretative map. This map will be presented in Chapter 4 (p.60).

We also consider the images from the short films and photographs to be relevant, as they help to revive the memory of what was experienced as carto-facts, and thus identify places and spatial references in the village, its geosymbols, and the itineraries of the parades and festivities in reverence to St. Sebastian and St. Benedict (SEEMMANN, 2008, HOLZER & HOLZER, 2006). This *interpretation* is based on the hermeneutic phenomenological approach of the anthropologist Cliffort Geertz (1989, 2002). Through the map, we tried to represent the village's ecology as Khôra, a space in movement that doesn't allow itself to be captured by simple geometry. This term comes from Plato (427-347 BC) in TIMEU (2001), and was appropriated by Augustin Berque (2000) to illustrate a social space that is the mark and matrix of genesis, meaning that there is an indissoluble ontological relationship between places and (social) things in the sensible world. Chôra for Berque (ibid.) is an occupied extension whose limits are determined by the occupation itself; this occupation takes place through the advance and retreat of something over another, and not a void to be filled .[14]

In this interpretive map we have tried to show both the route of the procession that brings the image of Saint Benedict across the River Itaûnas for the festivities, as well as the route of the procession of Saint Sebastian that goes through the whole village, returning to the parish church. The place where the congos meet to play Ticumbi and Reis de Boi, and where the Alardo[15] is performed in honor of Saint Sebastian and Saint Benedict.

We also represent the home of the main Ticumbi festival-goers. A festival-goer[16] is a social actor who invests in the festival, either financially or by donating a pig or ox for the raffle. They also contribute to the redistribution of food and accommodation for the various groups from other towns, including groups of women who play the jongo, and who come to take part in the festivities in the town of Itaûnas. There are three festival-goers in Itaûnas: César from Pousada Vila Tânia - festival-goer of Ticumbi de Itaûnas, Maria Inês - festival-goer of Ticumbi Santa Clara (from Angelim); Dona Maria Catarina - festival-goer of Ticumbi do Bongado (the oldest in the village). Although these are the official festival-goers, who contribute heavily, in fact everyone in the community contributes in some way as a festival-goer, especially the women (with their invisible work) who take care of making the clothes, feeding, arranging and decorating, as well as serving the guests. The tasks are distributed to everyone and, as far as I can tell, there is already an organized group in the village that is in charge of this organization.

However, it is necessary to think about these spatial configurations in constant movement, constituting several different landscapes throughout the year and especially on the days of the festivities: January 18, 19 and 20.

In interpreting the fieldwork, we sought to understand the spatial configurations and their formations woven in and through everyday life. This is intended to recognize and represent the landscapes of fixes and flows that form and (re)configure each social moment, as well as the social network of affection[17] and other networks of social relationships woven into the daily lives of social actors in a state of community .[18]

---

14 An empty glass is always full of air, as in Gilberto Gil's song, alluding to Heidegger's thought. An example of Khôra could be that stretch between the advance and retreat of the sea.

15 A medieval reenactment, a fight between Christians (Saint Sebastian - red) and Moors (Saint Benedict - blue), during the town's festivities, as we'll see below.

16 The *festeiro* category is very well described in Zaluar (1983, p.69), when he informs us of its origins in Portuguese culture in the custom of distributing food to the poor, especially during the Espirito Santo festival, and that in Brazil these elements of cultural heritage have constituted their own symbolic means of representing prevailing norms and values.

17 On this subject, see the article by Dias (2007, p. 11-24) and Scherer-Warren (2007, p.27-50). We'll discuss this further below.

18 Moscovici (1990) speaks of a 'state of community' (as a feeling of belonging), as well as a 'state of multitude' (anomie).

1.3 Geographical location.

The Vila de Itaûnas district (figure 3) is located in the north of the municipality of Conceiçao da Barra, ES, in an area of coastal plain, of Quaternary deposition, between wetlands. The district has an estimated population of 2,800 inhabitants - rural and urban (FERREIRA, 2002). The town (urban area) has an estimated population of 1,500, according to the Institutional Diagnosis of the PEI-2000, and in the high summer season it receives around 60,000 to 70,000 tourists. In the state's geographical context, the municipality of Conceiçao da Barra, ES, is located in region 6, on the north coast of the state of Espirito Santo, bordering the state of Bahia, as shown in figure 4.

The particularity of the village is that part of its territory has been (re) defined as a Conservation Unit of restricted use, as we *have already* seen, the Itaúnas State Park (PEI, 2006).

The village is made up of three localities, according to the PEI (2006):

1-Riacho Doce (bordering Bahia) - 75 inhabitants

2-Paulo Vinha (a rural settlement) - 350 inhabitants

3-Angelim (a territory of remnants of aquilombados) - 70 inhabitants.

The district has an area of great scenic beauty with dunes, beaches, a river, wetlands and fragments of Atlantic Forest, as shown in Figures 2 and 3.

Figura 2: districts of Conceiçao da Barra, ES. Source: courtesy of INCAPER, 2007.

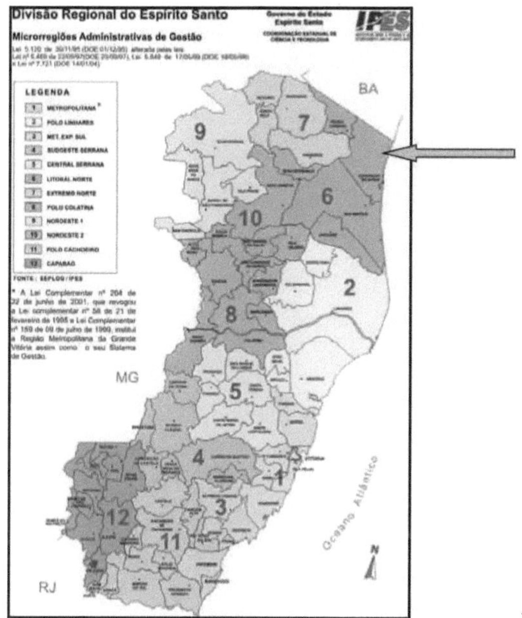

Figura 3: representation of the geographic territorial context of the municipality of Conceiçao da Barra, ES (source: IPES, 2000).

Figura 4: view of an area of Vila de Itaùnas, with the River Itaùnas and the wetlands in the background. Source: Photo provided by the PEI Manager, 2008.

Figura 5: another view of Vila de Itaùnas in the background, the Itaùnas State Park, the fragments of Atlantic Forest, the Itaùnas River, the wetlands surrounding the river, the dunes (left side), the beach and the eucalyptus groves in the background (Aracruz Celulose). Source: Photos provided by the PEI manager, 2008.

1.4 Presentation and development.

Each chapter has been constructed with a view to understanding the issue of spatiality and its expressions, territoriality and identity processes, power games, institutions and social networks, the moments of the conniving landscape (as a snapshot of a moment, an image captured and interpreted), all in a dialectical process. In the first part of this book, we present the introduction itself, the methodological path and the geographical location of the town of Itaúnas. We then situate the town over time/space, and analyze the processes of its formation with the chapter on the socio-historical-geographical-cultural context, with its various events.

In the next chapter we discuss the concepts we work with, such as spatiality, territoriality, place and identity, articulating them. We then present the different moments of the conniving landscape in the village of Itaûnas: mark and matrix, fixed and flowing, with the representation of the Ticumbi festivities and parades in honor of the Catholic saints. In order to understand how this festival is organized, sustained and represented as a great rite of passage, we will talk about the social networks, woven into everyday life and maintained by the affectivity of the participants, in the reciprocity[19] of Mauss (1974), as a moral debt to the saints in a demonstration of faith. As well as the category of *festeiros* and *festeiros obreiros*, as the nodes in the network, those who sustain this ritual of faith and joy. Finally, in the sixth chapter, we present an ethnographic work, adapted to geography, and therefore ethnogeographic: making with healing knowledge, properly speaking, as an expression of this differentiated spatiality, this *modus vivendi,* a *society-nature relationship.* We clarify that in this chapter we use the first person singular (I) and plural (we), because of the interpretivist method[20] , a cultural analysis according to Geertz (1989).

In the Vila we find the specialists in the art of healing, the official healers, midwives and pray-ers, in their social place in the institutions, as representatives of a suborder of Brazilian popular Catholicism, devotional, non-clerical. We will discuss this socio-historical form of Catholicism and its transformations, demonstrating how these practices are reproduced with knowledge, their corpus of knowledge supported by the empirical experience of what is lived and perceived, as well as by the intertwining of the cultural knowledge of white Europeans, black Africans and Indians.

As the village is next to the Itaúnas State Park (PEI), and there is a conflict between the management of the PEI and some of the community, we will talk about this conflict, but we won't go into it in depth given the limits of this work. So we'll work on this theme in a very diluted way, so as not to lose the focus of the work, and in this case at some point this issue is addressed, but never as a central theme, but as a supporting one.

---

19 The gift, or reciprocity, which involves the threefold: giving, receiving and giving back (MAUSS, 1974).

20 The interpretation, as already explained in the method, passes through the perception of this observer who went to the field, listened and analyzed the speeches, gestures and acts of the local actors, who in turn are also interpreters of their culture. In these terms, the ethnographer is an author and the interpretation is a "construction of other people's constructions, of what they and their compatriots propose" (GEERTZ, 1989, p. 19). At other times 'we' was used, showing that there was more than an interpretation, a reflection supported by the authors who study the subject.

2. Socio-historical-geographical-cultural context.

According to Ferreira (2002, p. 15), the word *Itaùnas* has its origins in "ita c Y-ta, what is hard, stone, crag, rock, pebble, metal in general, iron", and "una, adjective negro, black, dark." In the words of the interlocutors, Itaùnas means "black stone" that exists along the bed of the Itaùnas River and gives its waters a dark color. The town takes its name from the river.

This stretch of the far north of Espirito Santo is part of the coastal plain of Quaternary deposition that begins at the height of the Doce River, characterized by the geomorphological feature of fixed dunes and surrounded by trays of Tertiary sediments. The original vegetation is tropical rainforest, spread over the Tabuleiro Forest, the Alagado Forest, the Restinga Forest, the Dunes and the Mangrove Forest. The old town was located on the sandbank between the River Itaùnas and the sea. For some reason, and according to various legends, the original vegetation was removed from this stretch. With the loose sand and the action of the northeast and southeast winds, Vila Antiga began to be buried in the 1950s, according to Ferreira (2002). However, Mota et al. (1998) points to the probable start of the burial in 1925 and the partial burial of the village in 1956. In 1942, the village became the subject of an investigation because of the burial, and in 1949 the Riacho Doce Forest Reserve was created in the district of Itaúnas, on the border between Espírito Santo and Bahia. In 1960, the new town of Itaùnas was built, and the old town was buried in the 70s. The burial "event" leaves as its legacy the mobile dunes, the flooded areas and the current town (FERREIRA, op.cit.) with its streets.

According to Ferreira (op.cit.), the sea is open and has no inlets or bays. The black rocks that protrude from the shore act as a natural reef and provide a more favorable place for fishing canoes to enter and exit. Interviews with fishermen confirm this information, and this natural reef also appears as a "fishing ground", an attraction for some species that hide among the rocks. It is known that the entire region of the Sâo Mateus River was a stronghold of the Tupinikin ethnic group (Tupi people, cf. JOGAIB, 2005) with warrior habits, who lived by hunting, fishing and farming, practicing navigation and occupying an extensive area of what is now Brazilian territory. The Botocudos occupied a large part of the Rio Doce and offered great resistance to the whites from the 16th to the 19th century.

The history of the place that became Vila de Itaùnas can be understood in four stages: the first of coastal fishermen and gatherers who used chipped and polished stone; the second period of agriculture, fishing and ceramics; and the third period of contact between the first Europeans and the indigenous population and the founding of Vila de Itaùnas (old); with the fourth the development of Vila de Itaùnas (old) (PEROTA, 1986, p.6, apud. FERREIRA, 2002, p.25). It is possible to say that we are now in the fifth phase of the village, with the arrival of international tourism, some foreigners as traders, as well as the strong crisis between a segment of the community and the management of the PEI, which we will discuss later.

The first village (velha in local parlance) was formed between the River Itaùnas and the sea, on former land that belonged to the Fazenda Itaùnas. This farm belonged to the ombudsman of the district of Porto Seguro, Mr. Marcelino da Cunha, and there lived Indians and blacks who took care of animal husbandry.

"Itaùnas is a livestock farm, with a corral or paddock for the cattle, and a miserable hut for the blacks and Indians who look after the animals. The owner had gathered a few families of Indians there to form a colony over time; they were initially intended to protect the coast against the Tapuias and Itaùnas, which is why [it] was considered a barracks. Some Indians, who happened to be on the same road as us, accompanied us north from Itaùnas. They were carrying hunting rifles and knew the area perfectly. We passed between two small streams, the Doce creek and the Ostras river, both insignificant, but which, coming out of a picturesque setting of verdant forest topped with beautiful palm trees, formed a romantic landscape" (MAXIMILIANO, 1989, p.173, apud. FERREIRA, 2002, p.30).

The description of the traditional customs[21] of the jongo, the congo, artisanal fishing and itinerant farming dates back to the 19th century, when Prince Maximilian described them on a trip to the mouth of the Itaùnas River, near the Guaxindiba mangrove swamp and now the town of Conceiçao da Barra (FERREIRA, op.cit., p. 28).

Jogaib (2005) reminds us that in the 19th century, the north of ES, precisely the port of Sao Mateus, was a recipient of black slaves, who were redistributed by trade to the district of Porto Seguro, the province of Vitória and Rio de Janeiro. In these terms, there were many black slaves in the region of Sao Mateus, Conceiçao da Barra and the surrounding area. Many of these "quilombos" produced manioc flour, and with the abolition of slavery in 1888, free blacks from the region of Sao Mateus and Conceiçao da Barra went to these existing quilombos and engaged in increasing the production of this flour. They even began to rival the region's large landowners in the production and trade of this flour. These settlements of aquilombados, even though free, began to suffer persecution from the landowners, including military force, claiming possession of the land. According to Jocaib (op.cit.), these populations then closed themselves off into small, almost self-sufficient communities in order to protect themselves. We can think of the cultural preservation of the Ticumbi manifestation (the name given to Conceiçao da Barra's distinctive congo), as well as the many "games[22]and African-rooted folklore in the region. This "closure" is probably due to the lack of cultural porosity in response to the intense threats to their territoriality. It's worth pointing out that these aquilombados communities have never stopped fighting for the right to their territory and territoriality, as the work of Ferreira (op.cit.) and Jogaib (op.cit.) shows us; even today, in the demarcation of the quilombo remnants of Linharinho (neighboring Vila de Itaûnas), for example, an intense legal conflict has been waged in the courts and in other everyday social instances .[23]

In the context of the 1940s, Ferreira (op.cit) presents the village of Itaûnas with around 340 houses, a mixed population with a second generation of European and indigenous immigrants, but still with strong Afro-Brazilian ethnic and cultural traits. The whole region around the village was aquilomb territory, with heavy logging as well. The village appears in the memory of "the old people" as a place with a busy commercial center and therefore much sought after by the surrounding population, with manioc flour being its main commodity. Its port was frequented by large canoes with a capacity of up to 60 bags that crossed the river to meet the trucks with buyers from Conceiçao da Barra and the capital Vitória. Another alternative route was to

21 Understanding tradition as a recurring social practice, as a way of dealing with time and space. On this, see Guiddens (1991) and Hobsbwan (2002).

22 Playing is what they say: we don't dance, we don't play, we play.

23 The subject is currently being reported on the website of the NGO Observatòrio Quilombola: http://www.koinonia.org.br/oq/noticias

take the goods (flour, rice, beans, corn, cassava, sugar cane and vegetables in general) in the canoes to Conceiçao da Barra, down the river Itaûnas (FERREIRA, op.cit.). These facts were widely reaffirmed in the interviews with the oldest residents, those who were part of the old town.

The village has always had a strong religious atmosphere, marked by its religious festivals, the main ones being those of St. Sebastian and St. Benedict (January 19th and 20th). Afro-Brazilian religiosity reinforces popular Catholicism with Catholic religiosity, with the imbrication of the sacred and the profane as a preponderant factor, as found in general throughout Brazil .[24]

In this sense and in this context, under the aegis of popular Catholicism, we can infer that Ticumbi in Vila de Itaûnas, this gathering of congos to commemorate the saints (Sao Sebastiao and Sao Benedito), is in fact much more than a "game", in the reference of the local actors. Ticumbi is seen in this book as an institution in the sense of Berger and Luckmann (1985), configured in the daily life of the village, legitimized in its actions, with its actors occupying roles and reproducing them through their strategies and tactics, creating a local cosmogony.

The festival takes place over three days, with various congos performing[25] ; various congo groups from the surrounding area, other towns and cities, including Vitoria, are invited to take part. There is also a staging of a medieval battle called "Alardo[26]with all the well-crafted and chanted music, the colorful head and body adornments, in a choreographed staging of great beauty, a manifestation of joy and faith - sacred and profane.

However, this is not just a festival with its ritualistic culmination in three days, since the community prepares for it throughout the year, with very well timed rehearsals, meetings to discuss the themes, the making of clothes and hats adorned with flowers and ribbons in contrasting, cheerful tones and an artistic sophistication of great beauty, as well as the organization of raffles, bingos and donations from traders and supporters[27] . These facts together gave me the perception that Ticumbi is the great

Religious institution in the town of Itaùnas.

In the old town, the interlocutors pointed out two streets, with a third being a passageway from one to the other. One community member, aged 71, said in interviews that there was a fourth street[28] that "would lead to the church" (certainly the second church rebuilt after the first one was buried by sand). According to the people interviewed in 2007/2008, there was Rua de Cima (near the sea) and Rua de Baixo (near the River Itaùnas). The wealthier residents lived in Rua de Cima and the less well-off in Rua de Baixo, as well as the commercial houses. A third street would be a passageway between Rua de Baixo and Rua de Cima.

With the slow burial of the old town of Itaùnas, the landscape changed, as the aforementioned author reports:

---

24 On this, see the works of Zaluar (1983), Steil (2001), Rosendahl (2008 a,b), Corrêa ((2004, 2008)

25 Ticumbi is the name given to a type of congo from C. da Barra, with a different rhythm and ritual, according to Rogério Medeiros in personal communication. When the actors say, "There's going to be Ticumbi", you have to understand that it's going to be a congo game.

26 We'll explain a little more about this staging later.

This is a manifestation of solidarity that takes place with other sympathizers who are not from the Vila, but who collaborate because of their faith, or for other reasons and interests.

28 It may not be exactly a street, but a 'path' to the church that was rebuilt after the first church was buried.

"Around this time, the old town of Itaùnas began to be buried with sand brought from the beach by the north-east and south winds. It took around 30 years for the sand to be completely buried, and during this time the residents tried to remove it from the streets, squares and houses in droves. As the sand came up against the obstacles of buildings and vegetation, it temporarily settled and built up mobile dunes in the old village area. Stories to explain the cause of this phenomenon? Many. One of them attributes it to a punishment from the patron saint Sao Bras, for being replaced by Sao Sebastiao. Others say that the cause lies in the removal of the restinga forest, which used to hold the sand in this coastal region with its geomorphological feature of fixed dunes. For what reasons was the vegetation removed?

This was due to the stench that emanated from the area used as a toilet; the danger posed by the shots fired by hunters around the village; and the access trails to the sea, which could be corridors for the winds to carry loose sand from the paths. Although there is no consensus among the possible causes, one element seems to permeate most of the versions: for some reason, the coastal vegetation in front of the village was removed, leaving the sand loose for the action of the north-east and south winds." (FERREIRA, 2002, p 36).

For Eliade (2001), every society needs an explanation for its cosmos[29] , which gives order to the chaos, its world, and this is also true of the Vila de Itaûnas community in relation to the burial. In the interviews, the residents add other myths to those exemplified above, such as the myth of the bug hole, as shown below:

I: Oh, there are various legends they tell. They say that there was... I've heard a story, right? that they told me that the sand came from a hole. Isn't there the beach of the hole?

E: Where is this beach?

I: It's a bit... About two kilometers from the beach here in Itaûnas.

I: Going to Conceiçao da Barra?

I: No, going to Bahia, to Riacho Doce. Two or three kilometers or so.

E: Passing Riacho Doce?

I: No, before Riacho Doce. People count... There's a hole there, right. A huge hole in the middle of the dunes there, which also have dunes. And the dunes there are older than the dunes of Itaúnas. There's a huge hole there that people say belonged to an animal. Fishermen who went fishing there said they saw a bug digging there. The people say that a lot of the sand came from there, and that the bug threw it to Itaûnas.

I: Wow, but it was such a big animal!

I: That's a story people tell. I don't know now. I don't know what kind of animal it is.

E: And the other story?

I: I knew more stories, but now I can't remember. A lot of people say that it's for Sao Braz, because in the old town... About a hundred and fifty years ago, the patron saint was Saint Braz.

I: Oh yeah? Wasn't it Saint...Sebastian?

I: No, it wasn't Sao Sebastiao. It was because of prejudice, these things, because Sao Braz was more... the ones

29 Mircea Eliade (1992), in his work "The sacred and the profane"

who had more devotion were the blacks, who had devotion to Sao Braz. Then people say that they swapped him for Sao Sebastiao, and Sao Braz put a plague on the fishermen's village here. That's another story, right, that people tell. Let me see if I can remember any more... there's also the story of the Córrego da Viraçao. Viraçao is the place up there. On the Itaûnas river, right up there. It's where a man called Duca Tora used to live. People say that he was a sorcerer, right, that he dealt with these things. It seems that there was a forró there, then a policeman came from there, there was a fight, then the policeman shot his son. Then he cursed that Itaûnas would never move forward again. Then there's another story that people tell... too. Wow, there are several that if we try to remember, I can't, I don't even remember... it's only afterwards, right, that we remember.

(E: interviewer; I: interlocutor, male, 16 years old, community; Vila de Itaûnas, C.da Barra, 2007)

Fonte: MAPLAN - Levantamento Aerofotográfico - Escala 1:8000 - 06.03.1997

Figure 6: View of an area photo of Vila de Itaûnas, with the River Itaûnas and the wetlands in the background. Source: Ferreira (2002, p.18)

As we have already seen, as an alternative, another village was built on the other side of the Itaûnas River, on an old farm owned by Mr. Lauro Vasconcelos, Seu Dodô, bought by the Conceiçao da Barra town hall. The plots were donated by the town hall, and the community members rebuilt their houses. Part of the demolition was used to rebuild the (new) Vila de Itaûnas, such as roof tiles and wood.

Of the old village, only the church building remained with its rocks and bricks bonded with mortar made from clay and shells, as a historical testimony, or roughness in the words of Milton Santos. According to Ferreira (2002), the village of Itaûnas today has around 2,800 inhabitants (see table 1), taking into account the rural and urban areas. Most of them are still in the "sertôes", or rural areas further away from the town, and include the communities of Assentamento Paulo Vinhas, Angelim (a quilombo remnant), and Riacho Doce. The urban area of the village, according to the PEI's Institutional Diagnosis (2006), currently has around 1,500 inhabitants.

According to Ferreira (op.cit.), part of the community in the district of Itaùnas is extractivist, especially those who live in the "sertòes", with fragments of wet tropical forest and the sea as the basis of their survival. However, in the village, although artisanal fishing is still carried out today by men[30] who regionalize it according to productivity, techniques used and species found, they no longer survive exclusively on it. The "livelihood", as the community members put it, of most families in the village comes from the tourist trade, whether it's renting their own house, renting the many "suites" built in the backyard of the houses, or working in the many inns, bars and restaurants in the area, as a bricklayer, housekeeper and caretaker. Many families also supplement their budget with the Bolsa Familia and the "salà defeso" that they receive for three months (from November to February).

Figure 7: Typical residential building in Vila de Itaùnas.

In the words of the older community members, the process that produced the grief and loss of the "old" village boosted tourism and "developed" the "new" village. The landscape is made up of the River Itaùnas and its marshland, the dunes and the beaches. This exotic and bucolic scenery is now a landscape that attracts mainly young couples and families, who have a second holiday home in the village. Tourism has been on the rise since the 80s, with a big boost in the 90s. The first visitors came from São Paulo and Minas Gerais and opened the first inns and commercial establishments, albeit still timidly. In the 1990s, the village was heavily invaded by disorderly tourism, which was only curbed by the establishment of the Itaúnas State Park (PEI), which, together with the local community and local and non-local traders, created a certain order for the place.

Nowadays, the village is also receiving foreign tourists, mainly Europeans from Italy, Germany, England, Lugoslavia and Slovenia, as we saw during our fieldwork. In 2006, the Italians bought some properties in the village, built them and started competing in the commercial sector with hostels, bars, a nightclub (called Acropolis, but which didn't work out and is now a forró) and an ice-cream parlor. A mansion owned by an Italian family was built in the eastern part of the village, and the style of the residence stands out from the whole architectural ensemble of the village. The town's landscape is changing as a result of these interferences,

30 On the subject of fishing, we'll talk later about a moment when the women take part in pulling in a net. This collaborative practice is part of the network of affection and reciprocity and is called "Piem".

which appear as the global acting directly on the local, see figure...

Figure 8: the building of a family of "foreigners" in the village of Itaùnas, completely at odds with the local "landscape". Verticality (global) crossing horizontality (local).

Table 1: Calendar of social activities, tourism and festivities from 2006-2008.

	Months of the year											
ACTIVITY	**01**	**02**	**03**	**04**	**05**	**06**	**07**	**08**	**09**	**10**	**11**	**12**
Ox Kings Festival (high season tourism)												
Feast of St. Sebastian and St. Benedict (Ticumbi)												
Carnival (tourism)												
Holy Week (tourist)												
Forró Festival (low season tourism)												
Sea bass closure												
Closed season (all fish)												
Christmas (Reis de Boi)												
New Year (high season tourism)												

One of the Vila's strongest entertainment attractions, apart from the beach, is forró, including the organization of a Forró Festival that has been livening up the evenings in July. Today there are two official forró bars in the village: Bar Forró (the oldest, see fig. 10) and Forró Buraco do Tatu. In the meantime, an Italian opened a "nightclub" type establishment in the village, but for obvious reasons it didn't work out, so it reopened its doors as a "forró" in the summer in order to attract people. There are also many small bars where you can listen to Regaee, MPB and samba, including live music.

Figure 9: Partial view of the Forró Bar on the right, the first in the village of Itaúnas, and the Pedrolina Restaurant on the left.

Figure 10: view of a commercial enterprise run by "foreigners" in the village of Itaûnas.

Most of these commercial establishments, such as bars, are not owned by local residents, but by traders from outside. This is also true of the many hostels. This summer, 2007/2008, a new bar called Raizes opened in the village, where you can dance a good samba. There are numerous Pousadas scattered throughout the village and, in general, in the summer they are always full due to the turnover of tourists in the village. Some only operate their small businesses during the summer and long holidays, as well as during the Forró Festival in winter. In general, these traders don't really have a close relationship with the "place". On the other hand, there are those *outsiders* who live in the village, are traders, but have already started a family and share daily life with the community. However, they will always be outsiders in the eyes of the community. A closer look can see this in their words, in their conversations, in their daily relationships in the social space of the village. The "accents" are varied.

The expansion of Aracruz Celulose's eucalyptus plantations (set up in the 1970s) in the north of the state of Espirito Santo has had a major socio-environmental impact on the whole region and especially on Vila de Itaúnas, as Ferreira[31] (op.cit., p.147) shows in detail:

"The impacts pointed out include the company's 'liabilities' and more recent periods:

- conversion of diverse native forest into eucalyptus monoculture;
- attracting people from Bahia and Minas Gerais to work as laborers in the municipalities surrounding the forest, a process that generated a large mass of unemployed people;
- land acquisition through land grabbing mechanisms;
- unemployment of 4,000 workers in the north of Espirito Santo, due to the mechanization of logging;
- land conflicts with the Tupinikin and Guarani in the municipality of Aracruz,
- conflicts with black people in 21 quilombo communities and other traditional communities who now live "inside" the eucalyptus plantations;
- use of the company's forestry waste by charcoal factories, using child labor and poor working conditions;
- Forestry development with small producers;
- countless former forestry workers who have suffered accidents;
- eucalyptus plantations in areas of permanent preservation such as springs, decreeing the death of several watercourses, the production of drought and the water crisis;
- the power relationship established between the company's top echelons and state and municipal public bodies, influencing the definition of agricultural and environmental policies;
- co-option and corruption of forest workers' leaders in Espirito Santo and Bahia;
- lawsuits against NGO and trade union leaders;
- impacts on the road surface and traffic on the BR 101."

Under the above conditions, we found the village of Itaúnas at three different times during our fieldwork: December 2006, January and February and March 2007 and January 2008, with a population of around 2,800 inhabitants[32] . In the summer, according to the PEI manager, the village receives around 60,000 to 70,000 people. However, the Vila community does not have sanitary sewage services, nor sewage treatment, and its inhabitants still live with precarious garbage collection and heavy tourist bus traffic during the high summer

---

31 For details of the misappropriation of land in the north of ES by Aracruz Celulose, see the dissertation by FERREIRA, USP, 2002.

32 An estimate of my own and inspired by the data that Ferreira (200) sets out in his work.

periods.

This intense, disorganized tourist activity (and all the bad things it entails) has been on the increase in recent years and there is no master plan to support this activity and various aspects mentioned here go unchecked.

Figure 11: piled-up garbage, disorderly parking of cars and buses.

Given the rapid social change that has been taking place since the 1990s, community life has already been transformed and this is visible in its landscape every year. Late modernity or high modernity, with its compression of time and space (HARVEY, 2005), also represented by the entry of big capital through Aracruz Celulose, Petrobras, and especially the phenomenon of disorganized tourism, could definitively transform this "place", making it more of a tourist "space" without a "soul" (HILLMAN, 1993), as was the case in Porto Seguro (Bahia), Bùzios (RJ), Trindade (RJ), among others, to name but a few.

This work shows that this "place" called Vila Dunas de Itaûnas has a "soul" built up in the social/cultural relationships of reciprocity (giving-receiving and giving back) of its inhabitants in a 'state of community' (MOSCOVICI, 1990), also in their relationship with society/nature. This 'soul' is a constituent of and in the act of spatialization of humans and can be found in the various cultural expressions, including the art of healing; in the knowledge/doing of healing as a *corpus* that some specialists (benzedeiras, midwives, rezadores, erveiros) in the community keep in their memories. This oral art (with its prayers, rituals and techniques) may be being lost because it is not codified in writing, but also because it is inhibited by biomedical practices that disqualify it. Local practices about the health/disease/healing process are relevant, but there is no project or movement to recover this know-how, this art of healing.

3. Spatiality, territoriality, place and identity: some conceptual issues.

This chapter aims to broaden our understanding of the concepts of spatiality, territoriality, place and identity, and to articulate them. Within the limits of this work, I will work with these concepts, but within a process of *being-in-the-world*; where spatiality, territoriality and the constitution of a place are understood here as *moments of this being-in-the-world.* We will see how these processes are constituted and give rise to the identity of a group, indicating a communitarian way of doing things. As Sodré (p.22) puts it: "what gives a group its identity are the marks it imprints on the land, the trees, the rivers." In this sense, it was necessary to make an interpretative reflection with the authors called upon to contribute to this achievement.

3.1 Space and spatiality

What is this space? What space are we referring to? The space we are referring to is presented as in Santos (2004, p.63), when he says that space is formed by an inseparable, solidary and also contradictory set of systems of objects and systems of actions (immaterial, material and symbolic, dialectically), not considered in isolation, but as the unique framework in which history takes place. This space is not a geometric space, but a socially constructed space. In this sense, the German philosopher Heidegger (2002) tells us that man's being in the world is spatial. For Heidegger, space is *dwelling*. Man inhabits the world, worldizes the world, since he creates a world to live in, gives it meaning, organizes it. In the sense that this author lends, we can infer that man spatializes the world, creates directions, senses and meanings.

"To dwell, to be brought to the peace of a shelter, says: to remain pacified in the freedom of belonging, to safeguard each thing in its essence. *The fundamental feature of dwelling is this protection*. Safeguarding permeates dwelling in all its fullness. It shows itself as soon as we are ready to think that being a man consists of dwelling, and this in the sense of a de-morphing of the dwellings on this earth." (HEIDEGGER, 2002, p.129)

And Heidegger (ibid. p.165) goes on to say that man not only dwells, "but poetically man dwells", quoting a poetic text by Hoideriin. For Heidegger (op.cit.), poetry cannot be reduced to literature alone, because literature is only one form of poetry's existence: the literary form. When the poet speaks of dwelling, says Heidegger (op.cit.), he glimpses the fundamental trait of human presence, where he sees the poetic from the relationship with this dwelling, understood in its vigorous and essential form. He goes on to say that the dwelling place of being is language. In view of this phrase, it is necessary to look for another philosopher, Hans-Georg Gadamer (apud. SHUCK, 2007), a successor to Heidegger, who says that "being that can be understood is language". Because language becomes a means of expressing being in the world. For Gadamer's hermeneutics (apud. LOTT, 2007, p. 13), all forms of expression present themselves as language, but a language that transcends linguistics. It transcends explicit speaking - simple speaking, speaking in itself. This speaking can take place in the form of dialogue, texts, works of art or even through the 'inner word' of each individual.

When we reflect on what Gadamer (op.cit.) says, we realize that this *language will be the dasein* of Heidegger, existence, being-ai, being. This relationship of being with the world, with space, dwelling presents a quality of being-there, expressing existential being. Heidegger's idea of existence reveals that human subjectivity is not real without the world (HOLZER, 1997, p.39).

Without wanting to delve into this philosophical question, we think that it refers to how this being spatializes itself, how it geographizes the world. And the being does so poetically, since this human being puts *meaning* into this inhabiting, in this spatializing, the being lends its quality to the place. And by expressing his spatiality, he presents his ethos, his culture, his *habitus*[33] (as a disposition of the subjects).

This sense of *dwelling* goes beyond instrumental rationality and presents itself as a poetic sense, where man places his essence. Heidegger (op.cit. p.178) says: "It is poetry that allows man to inhabit his essence. Poetry allows us to dwell in an original sense." In other words, he constructs a sense, a meaning, to his being-in-the-world. This metaphysical animal, according to Schopenhauer (apud. DURANT, 1996, p.295) inaugurates a *place* with all its symbolism. In this sense, Eliade says (1996, p. 8):

"Symbolic thinking is not an exclusive area of the child, the poet or the unbalanced; it is consubstantial to the human being; it precedes language and discursive reason. Symbols reveal certain aspects of reality - the deepest aspects - that defy any other means of knowledge. Images, symbols and myths are not irresponsible creations of the psyche; they respond to a need and fulfill a function: to reveal the most secret modalities of being."

Porto-Gonçalves (2002, p. 2) makes an important observation regarding the geography of the social, which helps us to reflect on it when he says that: "We start from the assumption that there is no such thing as an a-geographical society, just as there is no such thing as an a-historical geographical space. Just as all geographical space is impregnated with historicity, history is always impregnated with geography."

And yet, "if history becomes geography, it is because, in some way, geography is a historical necessity." Well, we can put it another way, that if humans make history, then they historicize the world. If he historicizes the world, he also *geo-graphizes* the world, as Porto-Gonçalves puts it, by inscribing his history on it, creating a world, a cosmos[34] , in this sense all history is social and all geography is social in a way. In a further provocation, I would say that if geography is social, it is also cultural, since there is no society without culture (culture as an agreement, a worldview, an *ethos* in the view of GEERTZ, 1989).

Regarding spatiality and culture, Corrêa (2008, p.302-303) explains that spatiality is conceived from the point of view that the fact is that processes and phenomena, both natural and social, are differentially inscribed in terms of spatial distribution. In these terms, Corrêa continues, spatiality has led to new questions involving diverse spatial configurations, as well as symbolic (non-material) ones, referring to those spatial distributions identified by Bonnemaison (2002) *as geosymbols* loaded with meanings associated only with *place*. Spatiality also encompasses the paths taken by man, paths that do not find a logic in the relationship involving distance-time-cost, but logics established in the domain of symbolic values, impregnated with specific meanings (CORRÊA, ibid., p. 303).

33 As in Bourdieu (1997, p. 42): "the *habitus* is that kind of practical sense of what one must do in a given situation - what we call, in sport, the sense of the game, the art of anticipating the future of the game inscribed, in outline, in the current state of the game."

34 Cf. Berger and Luckman (1985), as well as Mircea Eliade (1992).

3.2 Territory and territoriality.

The author Sack (1986) presents territoriality as an action promoted by its agents in their relationship with the world, revealing a powerful geographical strategy with the aim of influencing and controlling people, phenomena and relationships, delimiting an area as a territory. These relationships are closely linked to power relations[35] as a strategic resource, a tactic[36] for inserting and controlling a particular social group. In fact, they are the means by which *space and society* are related in a given perceived historical time.

This approach, although sociogeographical in date, still limits action to "power" and Lopes de Souza (1995) goes further by proposing a more flexible notion of territory. For this author, territory is best represented as a "field of forces"[37] , a web or network of relationships which, in addition to its internal complexity, defines a *boundary* and an *otherness* at the same time. It thus establishes the difference between 'us', the group, the *ethic*[38] , the members of the collectivity or community, the *insiders*, and the 'others', the *outsiders*, the *ethic*, the strangers, the *outsiders*. In addition, Lopes de Souza says that territory is fundamentally "social relations projected onto space". We can infer from this way of thinking that this "relationship" or complex of "relationships" establishes the place, the proper, the particular of a given social group - its agents in relation to a space/time (Holzer, 1996).

The author Muniz Sodré (1988, p.12) works with the idea of territory as a space-place, something referenced by history and the memory of the agents who built it. He also shows us how the "being-in-the-world" of the human subject is spatial, in a reference to Heidegger, as shown above. For Sodré (ibid., p.23), the notion of territorialization is endowed with an active and symbolic force, a force that is demonstrated in a relational system between beings and objects, a *game* .[39]

"The idea of territory does indeed raise the question of identity, as it refers to the demarcation of a space in difference with others. Knowing exclusively or the pertinence of actions relating to a given group also implies locating it territorially. It is the territory that, in the manner of the Heideggerian *Raum*, draws boundaries, specifies the place and creates characteristics that will give shape to the subject's action. (...) Territory is thus the marked place of a *game*, which is understood in a broad sense as the protoform of any and all cultures: a system of rules of human movement of a group, a horizon of relationship with reality. Articulating mobility and rules on the basis of "make-believe", of a founding artifice that is repeated, the game appears as the ordered perspective of the connection between man and the world, capable of combining "the ideas of limits, freedom and invention"'.

In this sense, Santos (2004, p. 63) also works on this relational aspect of space, including the image of a hybrid,

---

35 See Foucault (1979) and, in the field of geography, Raffestin (1993).

36 See Sack (1986) and Certau (1990).

37 This notion of territory as a 'field of forces', a set of relations, goes hand in hand with that of Bourdieu (1989, p.134).

38 According to Geertz (2002, p. 87): *æmic* - refers to the local, a close experience; *ethical* - refers to the non-local, the other, a distant experience.

39 *This* game that Sodré (op.cit., p.23) refers to is *the* game that Gadamer (1985, p.38) talks about in another chapter; it's a game that goes beyond serving an immediate purpose, and in a very broad sense, it's a game, an unfolding of an original form, oscillating between calculation and risk, of the subject's relationship with reality. It is the game of self-movement, a basic characteristic of what is alive.

a system of objects and actions, a mixture of the material and the immaterial. Santos cites Godelier (1966, p. 254-255, *apud.* Santos, 2004, p.102) in bringing these ideas together: "every system and every structure must be approached as 'mixed' and contradictory realities of objects and relationships that cannot exist separately." In line with these concepts, we can infer that territoriality is a subjective and relational fact, which can be known and recognized through its various forms and expressions, as a cultural heritage. This heritage is presented as material or ideal in local knowledge. Without going overboard, we can also add that territoriality is not only an expression of power, but also a strategy and tactic that is strongly dependent on the relationships and positions of its agents.

In fact, they are a way of *being in the world,* representing an *ethos,* a *modus vivendi*, revealing a lifestyle and a cultural identity. These very plural forms or expressions have their own directions and create their own conformations, which do not necessarily obey a hegemonic, capitalist and globalizing logic of appropriation or (re)appropriation of space.

These spatial configurations can be seen in the itineraries of walkers on the "paths of faith" as presented by Sandra de Sa Cameiro (2008) in her studies of walkers on the Caminho da Luz, Minas Gerais; Passos de Anchieta, Espirito Santo, Caminho das Missôes, Rio Grande do Sul; and Rosendahl (2008a, 2008b) and Santos (2008) when they present pilgrimages to shrines and sacred places such as Fatima, Portugal, and the Camino de Campostela, Spain. These spatial configurations are also found in the village of Itaûnas, when we observe the Ticumbi procession or the procession of either Sao Benedito or Sao Sebastiao, since they inaugurate an ever-renewed, living and fluid spatial reference in the form of a *Landscape*, as we will see in another chapter.

These conformations can also be seen in the irregular design of a popular neighborhood like Maré and Rocinha, in the city of Rio de Janeiro. As well as in the landscape of a village or even a city such as Paraty, Rio de Janeiro, Sao Paulo, Arraial do Cabo, or even Vila de Itaûnas, ES. As well as in the process of configuring the internal and external spaces of a simple "house", a Place of dwelling in Heidegger (2001), as we saw earlier. In the language of Harvey (ibid., p. 196) and Certau (1996), these spatial configurations present real "trails of life", shown by Harvey (ibid.) as a diagrammatic representation.

3.3 Place

When it comes to understanding place, we will look to Tuan (1983, p.5) when he says that you can't define place without defining space (you can't define one without the other). In this sense, it is worth briefly recalling the conceptualization of space that is given here: space as socially constructed, a space of *dwelling* in Heidegger, the *Dasein* - the being-ai, existing in its daily life. It is in the sense of the experience of being in the world that Tuan (op.cit.) says that space allows movement, it is dynamic, and when the being lingers in a space, it gives it meaning, it inaugurates a place. In this way of understanding, Holzer (1997, p.71), when analyzing Buttimer's (1976) and Tuan's (op.cit.) thoughts on place, says that the content of places is the same as that of the *world* (in a Heideggerian sense), since both are produced by human consciousness and its intersubjective relationship with things and others. It should be clarified that in Buttimer's (1982, p.172) phenomenological view, the *world* is the context within which consciousness is revealed, far beyond the common sense world of facts and business, but a *world* of values, of goods. In other words, it is the world of symbolic goods, where a practical sense, a *habitus*, operates (BOURDIEU, 1997, p.42). In this sense, place is a privileged category in geography, says Holzer (1998). In this work, the privileged place of observation will be the town of Itaùnas and its social actors in a state of community.

3.4 Identity

The notion of identity can be discussed from various angles, but here we will focus on cultural studies in order to find a theoretical basis that can be articulated with and within the theme. As this is a rather recent topic, we looked for some authors who could favor this possible articulation. Among them, the most cited are: Silva (2000), Hall (2000a, 2000b), Woodward (2000), Moreira (2005) and Castells (2006).

The subject is relevant and topical given the destabilization of old identities - a phenomenon of late modernity, or high modernity, which for so long stabilized the social world. New identities are being constructed, and this modern individual (now on the edge) is somewhat 'fragmented' in this process, according to Hall (2000, 2000a). However, Harvey (2005, p. 219) says that, in the compression of time/space where the only conceived time is the present (the schizophrenic world) and space seems to shrink into a 'global village' of telecommunications and a 'spaceship earth' of interdependencies ranging from ecology to economics, we have to keep learning to deal with this overwhelming reality (an imposition) of the limit of Modernity. This experience presents itself as a great challenge to social/cultural identity, a permanent tension and also a stimulus. This disturbance has brought about a variety of social, cultural and political reactions. As spatial and temporal experiences are primary vehicles for the codification and reproduction of social relations, according to Harvey (ibid.) quoting Bourdieu, a change in these experiences varies representations and almost certainly generates some kind of modification in social relations, as in a circuit. Harvey (ibid.) also clearly demonstrates the influence of these representations of space/time on the identity of individuals. We will follow up on this thought later in our investigation into identity and difference.

In the view of Manuel Castells (2001, p. 22), the term identity refers to social actors and is constituted as an ongoing process of constructing a set of meanings, based on a cultural frame of reference, or what he calls "a set of interrelated cultural attributes, which prevail(s) over other sources of meaning". However, this plurality generates a certain tension and contradiction of self-representation in social action[40] . Because of this, Castells (ibid.) draws attention to the fact that multiple identities should not be confused with the roles and set of roles that social actors represent. These roles (worker, mother, neighbor, socialist activist, trade unionist, teacher, etc.) are defined by norms structured by the institutions and organizations of society[41] . Each of these roles has the capacity to influence the behaviour of the people around them and depends on negotiations and agreements between the individuals and the institutions and organizations. In this process, if social actors internalize their

40 According to Weber (2005, p.13) Social Action (including omission or tolerance) is guided by the behavior of others, whether this is past, present or expected in the future. The "others" can be individuals and acquaintances or an indeterminate multiplicity of completely unknown people. However, not every type of action (or external action) is "social action" because internal behavior is only social action when it is guided by the actions of others. (...) Not every type of contact between people has a social character, but only behavior that, in terms of meaning, is guided by the behavior of another person. For Weber, it's the meaningful connection between actions that matters.
41 See Berger and Luckman (2002) in "The social construction of reality".

role, it can become their identity.

As such, identities are sources of symbolic meaning for the actors and are originated and constructed by them in the process of individuation[42] , as well as in their relationship with others. Castells (ibid.) goes on to add that the construction of this identity, within this process (a dialectical game between I and we), draws on the raw material of multiple factors such as geography (I would say the strong connotation of the conniving landscape, questions of the construction of territo riality, place, etc.), history, biology and even productive and reproductive social institutions, through collective memory. In the psychoanalytic field, through personal fantasies (individuation), as well as through the apparatuses of power and important religious revelations (another symbolic power). All this baggage is processed by individuals, in a social or community state[43] , in social groups and societies that reorganize its meaning according to the tendencies, choices and cultural projects assumed in their social structure, as well as the vision of a time/space (perhaps a neologism fits: sociogeohistory).

It matters to Castells (op. cit.) who constructs the collective identity, and what this identity is constructed for, these are to a large extent the determinants of the symbolic content of this identity, as well as its meaning for those who identify with it or exclude themselves from it. This is why it is important to understand the interplay between the same and the different.

3.5 The game of same and different

The game of identity is established through and by difference, according to Silva (2000), Hall (2000b) and Woodward (2000). This game has a linguistic performance that is consolidated within a given culture, which devises classificatory systems for ordering a world. These acts of linguistic creation are understood as symbolic representations (in a mental/cultural way) that inform what is (the same) from what is not (the different). Culture is constructed (and is constructed) in a process between discourses, places (landscapes), and also the experiences of individuals in their social 'place', with themselves (psychoanalysis), in their family group, their school, work, in short with others and with the world (cf. BERGER AND LUCKMAN, 2002). From this I can infer that identity and difference, in agreement with the authors above, are not creatures of the natural world or of a transcendental world, but of a cultural and social world. The game of identity and difference is presented as socio-cultural creations that can be expressed through linguistic acts.

As language is a system of elements (signs), these have certain properties, which the play of identity and difference inherits or accompanies. Silva (2000, p. 77) argues that the signs in a language have no absolute value, and make no sense in isolation. These signs, roughly speaking, need to be connected to a chain or network of infinite events that are different from them (other graphic or phonetic marks). For example, I'm Brazilian, because I'm not Argentinian, nor French, nor Spanish. In order not to extend the subject, the author

42 This process has a psychoanalytic dimension and has been discussed by Woodward (2000, p. 60-67).
43 Cf. Weber (ibid. p. 141): "community can rest on all sorts of foundations, affective, emotional and iradditional (...)". However, for this author, there are no pure types, because the vast majority of social relationships are partly "community" and partly "society". Finally, for this author (ibid., p. 143) community requires a content of *meaning in social relations*, where a feeling of belonging, of a homogeneous situation gives rise to a *feeling of community*.

says that language (a system of signs) is nothing more than a system of differences. So I can say that difference and identity are co-producers of each other.

But Silva (2000) also says that the sign, this mark or sign, contained in a certain classificatory system, is vacillating. It vacillates because it stands in the place of something it represents, but it is not the thing itself. In this sense, the thing itself does not coincide with the concept. However, we have the illusion of seeing in the sign the presence of the thing itself. This is what Derrida (apud. Silva, 2000, p. 78), a priest of cultural studies, calls the "metaphysics of presence". For Silva, this illusion is necessary within the linguistic system for it to function, because the sign is always in the place of what it represents. It is itself a symbolic representation, part of a process and dependent on this game of differentiation. The sign bears the trace of what it replaces (it contains a meaning), but also of what distinguishes it, of difference. Sameness, or identity, always contains the trace of otherness, or difference (Silva, op.cit., p.79). The example of the dictionary is great[44] . When we consult a word, the dictionary doesn't present us with the thing itself or the concept itself, but refers us to other words, other signs that postpone the presence of the thing itself or its concept. In fact, it forms a general idea within a classificatory, linguistic system.

Silva (2000), inspired by Foucault, informs us that this classificatory system, where the sign, the language, is located, is a discursive symbolic production and thus coincides with the power relations that created it (they are at the heart of this relationship). This is because to divide is to classify and hierarchize. What hides this power relationship is the naturalization of these binary, asymmetrical, hierarchical oppositions. One can have a positive value, the other a negative one. The normalization of the process can make one true and the other false[45] . The positive one is normal, the other is something outside the classificatory parameter and will be seen as negative and devoid of value within the cultural system that classified it as such. In agreement with this author, it is possible to say that 'ethnic' and 'folk' will be the music, religion or food of the 'others'. When it comes to the healing art of minorities, it will be called 'popular' medicine, 'folk' medicine, 'rustic' medicine, mysticism, witchcraft, always an asymmetrical counterpoint, without legitimacy in relation to the official, normal, positive, expected, desirable and only medicine capable of promoting healing in a hegemonic society that wants it.

In this case, Silva (op.cit.) states that: questioning identity and difference as relations of power means problematizing the binarisms around which they have been organized, thus agreeing with the problematizing ideology of educator Paulo Freire (1999) .[46]

Identity is not fixed, although there is a tendency in the power game to fix and stabilize it, just as there is with language. In fact, fixation is a tendency, but also an impossibility, since it is the result of a socio-cultural process that is always being constructed and (de)constructed.

---

44 The digital dictionary Aurélio XXI (or in FERREIRA, 1999), presents the linguistic unit as containing a signifier and signified; linguistic sign. The acoustic image of a linguistic sign is not the spoken word (i.e. the material sound), but the psychic impression of that sound, according to Saussure; in common usage, however, the term sign often refers to the word.

45 Cf. this naturalization in Bourdieu (1989, p.157).

46 When he evokes praxis as the reflection and action of men on the world in order to transform it.

In the case of national identities, Silva (op.cit.) shows that it is extremely common to appeal to founding myths in order to naturalize and fix them. In this sense, we need to be very careful not to slip into "cultural essentialism". As we have already seen, identity and difference are social and cultural constructions within a context. A founding myth is an invented tradition, and refers to a moment in the past when some event, gesture, usually carried out by a providential figure, inaugurated the basis of the supposed "national identity." Just to refresh, it's good to remember that myths are narratives that have no moral commitment to truth or otherwise, because the role of myth is to give meaning, to order. What matters is the strength of the narrative in recreating a fact, event or gesture that links the national identity, in this case, to a feeling and affectivity that sustains its stability within a context in which it is inserted. This is the function of myth, of the mythical narrative. It's not good to forget that there are 'biological' arguments about the fragile nature of women, the indolence of blacks, mulattos and Indians, for example, which are just as cultural as the myth, in fact they are also constructed myths. They are and contain power games within a given context that suits the interested party, the arguer.

At the limits of modernity, or late modernity, there have been movements that tend to conspire against this fixed identity and make it more fluid to the point of "mixing", transgressing borders. Thinkers of cultural theory use metaphors to describe these processes such as hybridization, miscegenation, syncretism, transvestism, border culture, in short, anything that gives the idea of movement, fluidity and displacement. These metaphors seek to counter the tendency towards essentialization.

The anthropologist Laraia (2003, p. 106-108), in a discussion on cultural diffusion, uses a text by Ralph Linton as an example of this *hybridization of* Americans in relation to 'their' customs, clothing and eating habits. He shows that there is nothing natural about this, and the fact is that everything has become naturalized as belonging to what suits this cultural system in the process he calls *cultural diffusion*.

Is there anything more to be said? The language makes explicit the tension of the hybridity generated by and in globalization, denouncing the transgression of symbolic and political/territorial borders.

It is possible to draw attention to the fact that social actors, like the *bricoleur*[47] , have the task of constructing their identity with all the material and immaterial artifacts at their disposal. We believe that it is in this sense that Castells (2001) emphasizes that identity has a what, a who and a why.

To illustrate this, Silva (2005) presents in an article that a given social actor, living in a peripheral neighborhood called a "favela", in a given event (looking for a job, for example) may identify himself as living in another part of the city, already anticipating the discrimination he may face. Another example is the resident of Saco do Mamanguà, who, when asked what he called himself, said: "Now we're not *mamanguaenses* anymore, we're *caiçaras*". In Xavier's (2004) interpretation, he uses a new identity - like a *bricoleur* - recognized by and through the discourses of environmentalists and researchers, seeking to gain a certain political strength in this social game.

To explain this, Hall (2000, p. 109) says that identity has as much to do with the questions of "who we are" or "where we come from", but much more to do with the questions of "who we can become", "how we have been

47 In an allusion to Lévi-Strauss's figure of lineage, said Bauman (2005, p. 55).

represented" and how this representation affects the way we can represent ourselves. Identities arise from the narrativization of the self, of a fictional nature, but this does not diminish their discursive, material and symbolic/political effectiveness.

3.6 What does identity have to do with space, territory and territoriality?

We've already said that space gets its essence from places, from man's dwelling, from this spatial ordering, giving meaning and significance to every action, object, nature, natures, through his body, in his body, which is also a space to dwell in and express:

"Protecting the quadrature, saving the earth, welcoming the sky, waiting for the divine, accompanying mortals, this four-sided protection is the simple essence of dwelling. Things built with authenticity mark the essence by giving a home to that essence." (HEIDEGGER, 2002, p. 138)

As we have already seen, territoriality is a strong geographic strategic action, an action loaded with symbolic effect which, in the social game through the interactions of agents and their positions, has the effect of controlling people and places, or demarcating a place as one's own. In fact, by giving "place" the status of "home", for example, or by giving a landscape a unique and symbolic meaning, as Tuan (1983) and Certau (2003) show us.

Lopes Souza (1995) tells us that territory can be understood as a "field of forces", a *web of social relations*, which defines a *boundary, an alterity*, which will lead to the process of constructing identity. This identity is a game and a process marked by *difference,* in the relationship, in social action, with and for others, establishing the limits of what I am and am not, given that the game of interactions contains relationships of power and force, of social positions, including the very act of territorialization as part of this game. This act, a social action (in the Weberian sense) in which the very act of territorializing is located, can shape a *lifestyle*, a *collective ethos* that delineates "geo-graphies", in the view of Porto- Gonçalves (2001) and social topologies, in the view of Bourdieu (ibid.).

3.7 A possible articulation: intersubjectivity.

Through the essay by Ayres (2001, p. 4), we think we can give a tone, a philosophical and pragmatic sense to these categories that need to be thought of together. This author evokes Kant's metaphor of the dove as an interpretative (hermeneutic) possibility for understanding human intersubjectivity (and its consequences). Through this imminent philosophical reverie[48] , proposed by Ayres (op.cit.), it may be possible to (re)think the possibilities of including the concepts of relational subject (individual/self), identity constructed in the game with difference, territoriality, as realities to be worked on.

The light dove, while in its free flight, cuts through the air whose resistance it feels, could imagine that it would be even more successful in the vacuum.

For Ayres (op.cit., p.4) the dreamy flight of the dove can be understood as a strong metaphor for the identifying act, in the attribution of predicates to the different moments of experience that make us constitute our worlds

48 Very close to the theory of strangeness that Ued Maluf (2002) suggests. As well as his mosaic of non-trivial isomorphs.

and ourselves, in a reference to Heidegger (1995). According to Ayres (op.cit.), "it refers to the process of *constructing identities*, which indicates to us an inexorable dialectic of denying by constructing/constructing by denying, so difficult to explain in words and so clearly expressed in the Kantian metaphor".

Continuing with this author's dialectical interpretation, the place of the subject as a thinking being, who by attributing predicates to the world, identifies himself. We are interested in the experience of "resistance" that makes these other subjects emerge as real presences, because it is in the friction with the other (otherness) that I am constituting myself (ipseity). The free flight of the dove can be understood as human history itself, the resistance of the air will be the inexorable presence of the other, in the reference of Ayres' article (op.cit.).

In agreement with this author, and in line with what has already been discussed in Identity and Difference, there is no such thing as a pure individual subject, only the dream of individuality born of the lived experience of not being alone, and of always and immediately being the 'other of each' (RICOUER, 1999, apud. AYRES, op.cit.). If human beings only exist in relation to other beings, then human beings only exist in relation. We need to follow Ayres' path (op.cit.).

Individuality/self and *individuality/self* are two of Heidegger's distinctions, used in Ayres' explanation (op.cit) to understand what the subject is. Individuality/self holds the strong sense of the subject, of subjectivity. The I (of identity/I) refers to the ontic, to being, to the existential[49] ; the self refers to the ontological, to being, to the existential[50] . The existential is relational, it refers to life, to existence, and we will be talking about the subject when we talk about the identity-self. But the I of identity/self which, like the dove, is only realized in the experience of imagining flying in a vacuum. Although in the Kantian metaphor the dove dreams, it does so because it has a desire (and will) that leads it to long for the happening of resistance against its wings. The dream leads the dove to conceive and experience birds, wings, flights and thicknesses in space. It spatializes itself in order to be identity-self (ipseity), the existential - relational. Finally, the pragmatic aspect is that the dove's dream takes place in the very act of flying, the existential sense, hence the existential - existential dialectic, as Ayres shows us (op.cit., p. 5)

> "In this existential sense, emptiness would be its very impossibility of being a bird, its wings would be of no use in emptiness, but in an existential sense it is only while it maintains the dream of overcoming the resistance of the air that the dove continues to fly, that it continues to be a bird."

In this metaphor, the aforementioned author identifies that the subject can only be a relational being, including in its spatiality. It is in this sense that Ayres (op.cit.) proposes to think of this relational subject in the rural community of Vila de Itaûnas, in its territoriality, its spatiality, in its place.

In the movement of this Weberian social action, social agents compose designs and shapes. This is why the space of the *house* (of dwelling) and its symbologies, and of the houses as a whole, form a social network[51] (which will be shown in another chapter and in the interpretative map), in a movement of reciprocity carried out by its agents (trails as *pedestrian rhetoric* in HARVEY, op.cit.). In this work, this movement will be

49 That which exists, thing, matter, object. That which we suppose to exist, Aurélio, 21st century.
50 The fact of existing, living, experiencing. Aurelian 21st century.
51 As Harvey (2005) explains in chapter 13, when he calls Bourdieu and Certau into an interesting dialogue.

interpreted as an objective referential framework (as presented in HARVEY, op.cit., p. 203), as a "grid" of socio-cultural practices that are expressed in a space. Along these lines, Corrêa (2008, p.306) brings us the question of symbolic itineraries:

"Regular symbolic itineraries take place on predefined dates, festivals, commemorating a political action, a religious devotion or a local tradition. Irregular symbolic itineraries, which do not have a pre-defined date, nevertheless have routes that have been established by practice, indicating the strength of certain routes. (...) The spatiality of culture is marked by symbolic attributes, associated with political commemorations, faith and protests, among other manifestations.

As a way of understanding this, I can see these symbolic itineraries, constituted through the spatial configuration of the houses, in the Sao Benedito parades and the Sao Sebastiao processions, in the way fishermen cycle on their way to fish, the coming and going of people looking for a word from the benzedeira, the rezador and in the locations of the rehearsals *of* the Ticumbi groups. There is also a symbolic itinerary in the coming and going of the local actors in the succession of days of the week, in the variations of the months of the year, obeying practices that are already constituted and institutionalized (as presented in the introduction). This *socio-cultural* whole in movement forms Landscapes and Geographies[52] that can be mapped as *carto-facts*, in the words of Seemann (2006, 2008), in line with Tuan (1983) and Certau (2003) and, more recently, Werther and Selma Holzer (2006), among others. This will be the endeavor of this work, to present these carto-facts, these doings with healing knowledge, as expressions of a spatiality.

52 As presented by Porto-Gonçalves (op.cit., passim).

4. Different moments of the conniving landscape in Vila de Itaùnas: mark and matrix, fixed and flowing.

Landscape is one of the oldest concepts in geography and was one of the first themes developed by German geographers using the cultural dimension, which was incorporated into cultural geography in the 1920s by geographer Carl Sauer, from the Berkeley School (MELO, 2005). In France, there were important contributions that focused on the cultural aspects of geography, as Claval (2002) shows, citing Jean Brunhes and Pierre Deffontaines. Claval (2002) points out that cultural processes are inscribed in a space, which is why cultural facts are of such interest to geography. In this way of thinking, "the landscape is the mark and matrix of a culture, it contributes to the transfer, from one generation to the next, of knowledge, beliefs, dreams and social attitudes" (CLAVAL, ibid., p.146; BERQUE, 2004).

Côrrea (2004), in his work, presents the semiographies of a landscape, which he calls *the conniving* landscape of the Candomblé yard-territory in Bahia. The term conivent landscape was coined by Sautter and appropriated by Bonnemaison (2002, p.107; CÔRREA, 2004, p.86) to give an idea of a landscape that has *Visibilities* (materialities), but is also impregnated with *invisibilities* (the immaterial), an encounter between meanings and signifiers. It will thus be the "place of an encounter and an almost sensual emotion between men and the land" that is updated through the generations and where *roots and culture* are imbued with a particular meaning conferred by the place (BONNEMAISON, op.cit., ibid.). In this case, the territory will be the embodiment of this culture, which can express itself in and as a Landscape - a conniving landscape - with its fixes and flows - reciprocally its processions (flows), and its constructions (fixes), its itineraries, its geosymbols. The latter as cultural meanings of a given space-time, which inscribe territorial identities - marked by ethnicity - as a semiotic[53] , and which can act as a terrestrial verification of the myths that are both the source of cosmic powers and the foundations of social organization (CORREA, ibid., p.18).

In this chapter, I will focus on the imagery of the cultural landscapes of Vila de Itaùnas. The landscape is thought of as a flow of images, constantly changing, and dialectically an eternal return to the same[54] , but always dissimilar. The semiographies will be described as a social cartography. A reading of a possible 'place', of a field of observation, and for this very reason, without the possibility of seeing everything, but understanding the unity contained in the *momentum.*

In this sense, the landscape may be a fragment, but each person who observes a *momentum* of this landscape will have a unique possibility of experiencing the whole - the unity of culture. Cosgrove & Jackson (2003, p. 137) tell us about this landscape and its imagistic power:

"If the landscape comes to be considered a cultural image, "a pictorial means of representing or symbolizing everything that surrounds the human being, then it can be studied through various means and surfaces: through painting on canvas, writing on

53 Bonnemaison (2002, p.93) uses the category in a broad sense, without reference to the biological, and therefore regardless of the existence of common ancestors (real or hypothetical). For him, ethnicity is linked to an awareness of oneself as belonging to a culture, and (re)producing that culture. "It is in their midst that the sum of beliefs, rituals and practices that found the culture and allow groups to reproduce themselves is elaborated and perpetuated."
54 Thinking about Nietzsche - the eternal return of the same.

paper, images recorded on film, and even earth, stone, water and vegetation on the ground" (Cosgrove and Daniel, 1987). Each of these means reveals the meanings that human groups attribute to areas and places, and makes it possible to relate these meanings to other aspects and conditions of human existence."

In experiencing this *momentum*, through its imagistic power, transcendence can occur. In this way, it is possible to infer the transcendence of the *experience of art* described by the philosopher Gadamer[55] (1985), which goes beyond an *aesthetic experience*. For Gadamer (op.cit.) this transcendence occurs through the process of hermeneutic identity of the game.

The concept of landscape as a configuration of symbols and signs leads geography to look for methodologies that are more interpretative than morphological. Since there are no static totalities, but totalities in movement (a quote from Sartre, apud Santos, 2004, p. 118). In this case, the landscape of Vila de Itaùnas that I am presenting is a totality in the sense of Unity beyond the sum of the parts, of form and content, in a dialectical process. This totality of culture can be perceived in its different expressions, which semioticize the territory through the lived experience.

What is lived appears in the social networks of reciprocity and affection, in the gift of giving, receiving and giving back from healing specialists such as the benzedeiras, the benzedor, and the midwives[56] . The lived and perceived space will be better demonstrated in the mental map made by the teenager Lucas, presenting "another view" - a view from within the culture, a discourse on space, and his world references.

In the town's landscape we will find a strong religiosity, although not always visible, as in the surrounding landscape. A closer look, combined with a given festive season, can bring out these religious expressions. This religiosity presents itself as a hybrid of popular Catholicism[57] and with a strong influence from the Afro-Brazilian culture of Ticumbi. These expressions constitute a territory where the sacred manifests itself, where the territory itself is sacred. Rosendhal (2008, p.56) informs us that the religiosity and faith of Catholicism can be found in the temples, cemeteries, the small oratories by the side of the road, the itineraries taken by pilgrims and hikers represent, among other things, the visible means by which the territory is experienced and recognized as such.

In this chapter we will discuss the landscape of Vila de Itaùnas as a text, a dynamic configuration of symbols that were "read" through the fieldwork, in the speeches and actions, strategies and tactics of the interlocutors and at various times (DUNCAN, 2003; COSGROVE AND JACKSON, 2003). The privileged approach is part of what Claval (2003, p.15) has called the cultural approach in Geography, or if you prefer, Cultural Geography. In relation to the term, we should clarify that when we deal here with "culture" as a tool, a concept, it should be understood as in Geertz (2001, p. 215) where he says: "What is culture if it is not a consensus?"

Following Geertz's (1989) thought at another point, we can say that in a community with a religious *ethos*, as in the case of Vila de Itaùnas, culture will be the tone, character and quality of their life, their style and moral and aesthetic dispositions - their worldview, their *modus vivendi*. A consensus between the parties, a truly

55 Preferably in "The Actuality of Beauty: Art as Play, Symbol and Celebration".
56 Although childbirth is not a "disease" or a state of imbalance, it is a liminal state.
57 Which in itself is already a hybrid, as Steil (2001) argues.

well-ordered state of affairs to accommodate a certain lifestyle.

In these terms we can infer that society operates through culture, but that there is a dialectical process between society and the individual, and in this sense we can agree with Elias, when he says that we are a society of individuals. And if we are a society of individuals, it is as individuals and society that we spatialize, semiograph the world, making it our "home". In the act of spatializing the world, we are also expressing our way of being in the world and with the world, through a world. In these terms, we create a "landscape" - a mark that expresses this agent's way of being in the world, and also a matrix, since in creating or producing a mark, we are also a product of this same mark. As in the words of Godelier (1990, p.17) when he says that human beings are not content to live in society, but that they produce a society to live in. In this case, the landscape can be seen as both a mark and a matrix, as Berque (2004) sees it. A mark of a society, a reference matrix of a space/time that remains in the memory of its agents who always reproduce it in a rereading.

3.8 Devotional festivities: fluid landscape, brand and matrix.

In popular Catholicism, the feasts of the saints are a moment apart, and they are traditionally conducted, planned and organized by the devotees. These festivals have a sacred character, and a seriousness and respect for the things of the saint, the many social meanings expressed through rituals and the very effectiveness traditionally attributed to festivals (ZALUAR, 1983, p.65). We also know that the feast is an extra-commonplace *moment*, a *moment in* a liminal ritual, which is why it is necessary to mark the times: the feast is a moratorium on everyday life, because celebrating feasts is a human act, according to Marquard (1998, p. 359360). Festive celebrations mark moments such as: wishing happiness to newborns, saying goodbye to the dying, wishing happiness to people getting married, giving thanks for good things that are not obvious, and finally, honoring God and the saints, God's mediators in the world. But the main thing is that man is a festive being, and therefore festive, and festivity is something exclusively human, in its symbolic and ritualistic sense. The human being needs the party: *"Living your life is the daily life of man; distancing yourself from your life, the party."* (MARQUARD, ibid., p.360).

As well as being a celebratory, devotional attitude, the festival has a strong identity factor, an integrator of the social body and therefore a regulator of social conflicts, a great catharsis of socio-spatial innovation (DI MÉO, 2001). The devotional festival, in this case, takes on the function of bringing the disjointed together, bringing the devotees into communion in the celebration, and demarcating a time/space as sacred. The devotional celebration provokes and evokes a sacred place - the experience of hierophany (ELIADE, 2001, ROSENDAHL, 2001, 2008 a, b, CORRÊA, 2008).

> Parades, processions and marches are the movements along symbolic itineraries. In reality, these routes have a double meaning. On the one hand, they give visibility to public demonstrations, and on the other, they demarcate certain spaces, temporarily transforming them into symbolic territory, creating an identity between the route and its content and a given group or institution. (CORRÊA, p.307)

We can say that the *complex ritual*[58] of the festivities of Sao Sebastiao and Sao Benedito (with all the added

58 Set of rituals or sequences of rituals, as unfolding events in which the same elements are combined (ALVES, 1980,p.30)

manifestations), in the village of Itaûnas, are landscapes that are formed, every moment a new image, a new scenery. This three-day social gathering is a festival of rhythms, sounds and colors, but that's not all. It presents itself as a cultural celebration of a religious nature - a marked encounter between sacred space/time and profane space/time (Eliade, 2001). In this sense, our intention is not to describe the details of the festivities, but to present the image as a representation of a landscape that marks the memory of social agents and contributes to cultural reproduction as a sign matrix. For Garcia (2001), these events teach young people about hierarchy, discipline and solidarity. This is why they are so important as a mark and matrix of a society with community relations (in Max Weber's sense of belonging). The entire route of the processions is shown on the interpretive map, see figure 29. In order to bring out the image, we're going to show the stages of the festivities, capturing some moments:

Figure 12: Arrival of the São Benedito procession by boat on the Itaùnas River.

Photo: Julio Silva (2008) kindly provided. piled up garbage, disorderly parking of cars and buses.

1- January 18th: the arrival of the image of St. Benedict, by canoe on the Itaùnas River, brought by the community from the Areia Branca site (the site of Maria Catarina, the Ticumbi do Bongado festival-goer).

2- The image is taken in procession to the church of São Benedito, which is not an official church, as it is not linked to the clergy, but is respected by all the Catholics in the village. The entire route along which the image passes is decorated, including the bridge.

Figure 13: image of Sao Sebastiao and Sao Benedito on the altar set up in front of the (unofficial) church of Sao Benedito, in a tent. Photo: Xavier (2008)

3- On January 19, the festivities begin with the Ticumbi games in honor of the saints. Profane and sacred are intertwined in a festival with a strong Afro-descendant cultural representation. Several Ticumbis perform in two tents. One is next to the Catholic church of Sao Sebastiao and the other, already mentioned, in front of the unofficial church of Sao Benedito, as can be seen on the interpretive map that follows the photos.

Figure.14: In the images, the representatives of the Ticumbi Institution are seated, while the others "play" in a procession in homage to the saints. Photo: Xavier (2008).

Figure 15: Staging of the Ticumbi do Bongado procession, with the Bamba and Congo kings in the background, and their secretaries and warriors to the side. Photo: Xavier (2008)

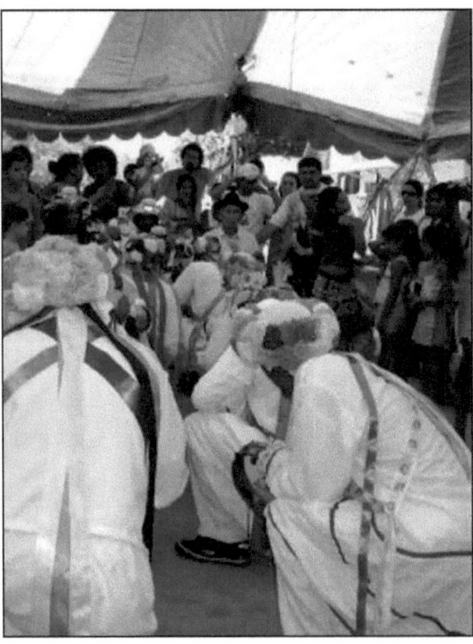

Figure 16: the choreography in devotion to the saints. Photo Xavier (2008)

Figure 17: The Ticumbi in a procession to Itaûnas State Park Headquarters. The PEI manager is on the right-hand side of the photo. Photo: Xavier (2008)

4- This is a moment when two institutions meet (see figure 17), two powers: the Ticumbi Institution and the Itaúnas State Park Institution. However, the PEI is recognized as a state institution, with powers to manage the environmental area of a restricted-use conservation unit, and with political intervention. The Ticumbi, on the other hand, are perceived and recognized by the state only as a "folklore group", with no political connotations, only a religious nature and no power to intervene. For these and other reasons, the invisibility and lack of political recognition of the community and its institutions has brought great harm, firstly to the community, and secondly to the environmental management of the PEI. There is a distancing between the community and the bodies representing the state, because of this invisibility that sustains an unequal power game. The state exerts its authority through force, the municipality through its representative, the mayor, sometimes sponsors some to the detriment of others, and the community experiences discontent that is sometimes veiled, sometimes made explicit in action, or denounced in banners in front of Praça da Vila, as shown in figure 18 below.

Figure 18: banner located in front of Praça da Vila, showing the discontent of the community over a dispute with the PEI. Photo:

Xavier (2007).

This banner remained in the same place until January 2008, when the fieldwork began. However, in this case, according to the interlocutors, the discontent does not represent the majority, but rather a group that feels discriminated against in a dispute with the PEI manager. The PEI, for its part, tries to maintain a policy of good neighborliness, but it seems that it has not been achieving its objectives for various reasons, as the manager claims in an interview. Some of the alleged reasons are the lack of infrastructure and economic resources for the headquarters, maintained by IEMA (State Institute for the Environment and Water Resources), the lack of trained human resources with graduates and postgraduates, as well as the mismatch between thought and action with some community members who, according to the manager, "act like children".

5- According to my interlocutors, the Alardo, also part of the festivities in honor of the Catholic saints, is a medieval reenactment made up of several acts and takes place over two consecutive days. In general, young people take an active part as soldiers. Older people occupy the positions of power and prominence, such as captain. The medieval struggle is represented by Moors and Christians, with the Moors wearing blue (represented by Saint Benedict) and the Christians red (represented by Saint Sebastian). The Moors capture the image of Saint Sebastian and the Christians try to rescue it. A real battle ensues with a sword fight, where the Christians win and recapture the image to put it back in the church. In the last act, on the second day, the Moors are converted into Christians and baptized. The whole performance is staged in the square in the town of Itaùnas, a central meeting point, as well as an observation point. Like the Ticumbi, this performance is richly detailed and there are rehearsals with the participants throughout the year. It was observed that the young people of the village take an active part in the festivities of Sao Sebastiao and Sao Benedito and this performance is one of their favorites. The other favorite is Reis de Boi, a syncretic performance in which the Reis festival is mixed with the traditional Brazilian festival of Boi Bumbâ. The staging is characterized by the cultural and ethnic intermingling of the people of Vila de Itaûnas (Indians, blacks and whites), but we can speculate that the geographical proximity to the Bahia border helps a lot. At the end of this Reis de Boi performance, the forest animals represented by puppets of alligators, dogs, the figure of the ox itself (boi bumbâ from Brazilian myths, festivals and legends), among others, frighten the young children who run around screaming in fear and joy at the same time. Perhaps this unusualness gives rise to joy and surprise, which is why it is preferred by children. The Reis de Boi da Vila also performs in these festivities in honor of the Catholic saints, and their performance is very popular. Believing in the power of imagery[59] , I present the festivities of São Sebastiao and São Benedito in the village of Itaûnas as *fact cards* (SEEMANN, 2008); their geosymbols, their itineraries, their spatial references represented in the figures below.

59 A picture is worth a thousand words.

Figure 19: Village square with the Mother Church (a geosymbol) in the background and the tent where the festivities in honor of Saint Sebastian, the patron saint, take place. Photo Xavier (2008).

Figure 20: Praça da Vila and its geosymbols, the pequi vinegar tree, the São Sebastiao flagpole and the parish church. Photo: Julio Silva (2007) kindly provided.

Figura 21: geosymbol that identifies the tent of the saint - Saint Sebastian being honored. In the form of a flag, it 'marks' a specific 'place' for the festival to *take* place. These semiographies are represented on the interpretive map. Photo: Xavier (2008).

Figura 22: geosymbol that identifies the tent of Saint Benedict, the honored saint. These semiographies mark out a space as sacred territory - hierophanies[60] , a place of enchantment and reverence, but also of joy and play: sacred and profane[61] are part of the manifestations of popular Catholicism. Photo: Xavier (2008).

---

60 Eliade (2001) presents the hierophany as a Figure 21: geosymbol that identifies the tent of the saint - St. Sebastian. In the form of a flag, it 'marks' a specific 'place' for the festival to *take place*. These semiographies are represented on the interpretive map. Photo: Xavier (2008).
manifestation that qualifies the sacred space.

61 This 'profane' refers to the Judeo-Christian ideology that presents this sacred-profane discrimination. In Afro-Brazilian sects, the guiding ideology is different and all of nature is seen as sacred (*natura naturans*), and all manifestations of reverence for saints, for example, are sacred. Dancing and singing are part of the ritual.

In popular Catholicism, especially in rural areas, the saints of devotion can be honored through novenas, litanies, festivals or dances. The *promesseiros*, as Zaluar (1983, p.58) tells us, use these social forms to pay *homage* to the saints. In this way, the more 'playful' saints, such as Sao Gonçalo and Sao Benedito, liked dances, while the more sombre ones stuck to novenas. In this sense, these expressions are religious, but also political (ROSENDAHL, 2001), since the games of position in the Ticumbi games are semi-graphing a territory in space, and affirming the identity of a group that establishes its territory/place. In this sense, we can see that territory and identity are linked, since territory favors the exercise of faith and the devotee's religious identity, and religiosity can only be maintained if its territoriality is preserved (ROSENDAHL,

2008 a, p.57). That's why this is the most anticipated festival of the year in the village of Itaûnas.

Figure 23: The unofficial church of São Benedito in the background, and a Reis de Boi play in honor of the saints. Photo: Xavier (2008)

Figure 24: official church in the background and a Reis de Boi game.

Figure 25: another angle of the Reis de Boi ritual in honor of Saint Benedict. Note the instruments and the head and body ornaments, as the body is also a reference in the semiotics of the religious/political institution. Photo Xavier (2008)

There is a very clear difference between uniform and costume, according to Da Matta (1973, p. 143-160). Fantasy allows for the ritual legitimization of everything that is tenuous, difficult to classify, of uncertain origin, which is located at a distance, but above all, which should not be openly acknowledged in everyday life.

Figure 26: performance by the Santo Expedito congo band, from Vitória-ES, a ritual in honor of the festivities of Saint Sebastian in the village of Itaùnas, with its symbols. Photo Xavier (2008)

Figure 27: ritual presentation of a Ticumbi *game* at the door of the Mother Church of Vila de Itaùnas. Photo Xavier (2008)

Figure 28: Ritual meeting of sacred and profane in the Mother Church. Photo Xavier (2008)

We can see the difference between the typical congo revelry of greater Vitória-ES and the Ticumbi *games of* northern ES. The costumes, the musicality, the intonation, the rhythm, the procession itself, the positions in the procession, the body ornaments and the musical instruments are visibly different, but Afro-descendant religious syncretism and the homage to Saint Benedict unite them in a network - a link of faith and tradition, a true immaterial and material cultural heritage, as we can see from the images.

These expressions of what is experienced present a fluid landscape, full of meanings and signifiers and in constant movement provided by its pedestrian flows, by the personal and social routes that leave their marks over the course of a space/time - their life trails[62] . To this extent, it is a conniving landscape, in the sense of Bonnemaison (2002) who configures it as a mixture of materiality and immateriality of a given culture embodied in its territory. Harvey (2005, p. 197), referring to De Certau, says that "walking" defines "spaces of enunciation" and this spatiality that I am presenting to you.

Their intersecting paths shape the spaces. They unite places, and thus create the city through the activities of daily movements. (...) the particular spaces of the city are created by a myriad of actions, all bearing the mark of human intention - pedestrian rhetoric.

Let's take a look through the iconographic material at the complex ritual in its various moments in the following pictures.

Figures: 29 and 30: "casaca" player, a typical congo instrument from the greater Vitória-ES region, and a set of congo instruments from greater Vitória, ES.Photo: Xavier (2008).

Figures 31 and 32: presentation of the Alardo (medieval battle) with the Christians represented in red and the Moors in blue. Photo Xavier (2008).

62 Thinking of Harvey (2005)

Figures 33 and 34: presentation of the jongo (danced by the women) and the Reis de Boi with the representation of the animals.

Photo Xavier (2008).

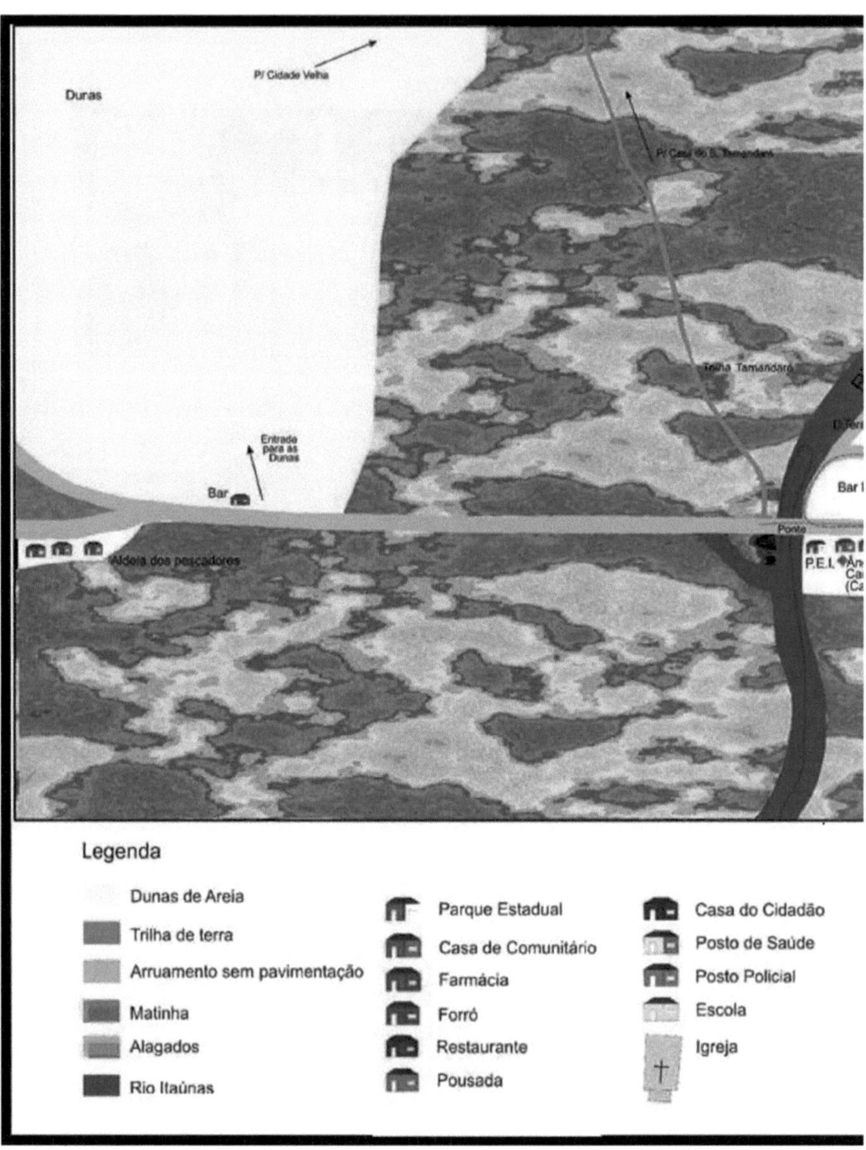
Duras
P/ Cidade Velha
Entrada para as Dunas
Bar
Aldeia dos pescadores
Trilha Tamandaré
Ponte
P.E.I.
Legenda
Dunas de Areia
Trilha de terra
Arruamento sem pavimentação
Matinha
Alagados
Rio Itaúnas
Parque Estadual
Casa de Comunitário
Farmácia
Forró
Restaurante
Pousada
Casa do Cidadão
Posto de Saúde
Posto Policial
Escola
Igreja

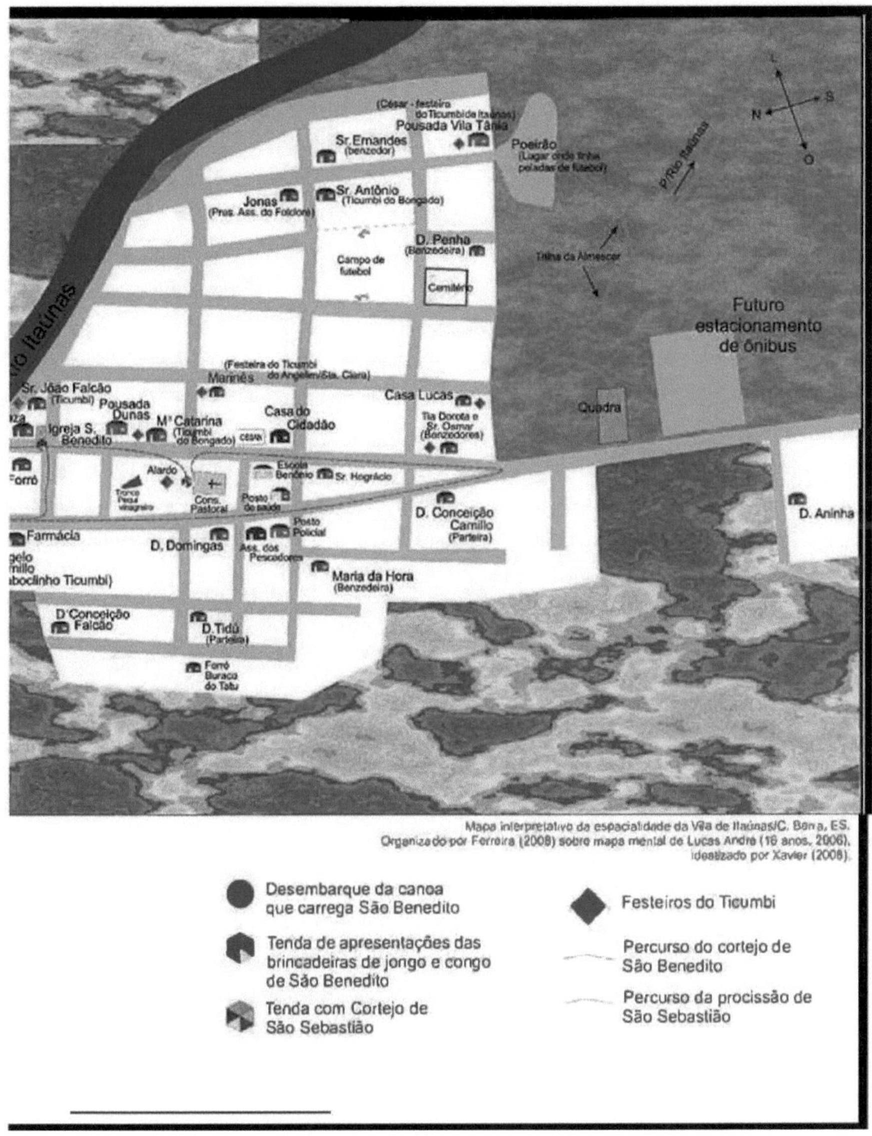

Figure 35(*) :63

4.2 The Soul of the Village

After this imaginary journey, we propose another to understand the soul of the town of Itaûnas. What is this

63 Interpretive map: the first basis was a mental map by the student of the "Benômio" School, Lucas André Maia dos Santos, 16 years old. This was digitized by geographer Renato G.S. Barcellos, and later worked on in Coreldraw by geographer Ferreira (2008). The expansion of this map was based on the interpretation and experience of the fieldwork,

"place" ?

In its early days, cultural geography emphasized the material elements of the landscape, privileging objects of study that were visible, tangible, in other words, where the difference appeared clearly, palpably, thus relegating other social, anthropological and even psychological dimensions of human existence to the background (SILVA, 2006, COSGROVE, 2003). These dimensions also configure a materiality, although not always visible, but recognizable in their expressions. Other social disciplines have made more progress than geography, Silva tells us (op.cit.). To this end, the study of cultural processes has become essential, since humans are tied to a web of meanings (GEERTZ, ibid) that they themselves have constructed, says Silva (op.cit.), in which Castoriadis (1985) agrees when he says that the essence of man is his capacity for creation, where cultural patterns are infinitely recreated in a cycle of the order of the eternal[64] . Holzer (2002, Cosgrove (2003) and also Porto- Gonçalves (2001), in another context, inform us that human beings experience and transform the natural world into a human world (geographize the earth); either through their direct and objective engagement as thinking beings, with their sensory and material reality, or as symbolic beings. The codes that human beings imprint are not only in language in its formal sense, but also in gesture, dress, personal and social conduct, music, painting, dance, ritual, ceremony and construction. For these authors, this list does not exhaust the series of symbolic productions through which humans maintain the lived world, since all human activity is both material and symbolic, production and communication (we are beings in dialogue with the world). This is why Santos (2004, p. 63) had already warned that "geographical space is a mixture, a hybrid, formed from the inseparable union of systems of objects and systems of actions".

In this case, culture operates not as a structure above humans, but as a set of control mechanisms, rules, plans and instructions for social interaction, a consensus, as Geertz (2001) puts it. Following this path of thought, we can say that the village of Itaûnas has a soul, often expressed in its landscape, present in the lived *momentum*, but also in memory - a conniving space[65] - which is what it presents itself as.

The soul of Vila de Itaùnas, the "place", is not visible to any traveler or tourist. The village keeps secrets behind each house, the dwelling place of the inhabitants, the "natives", as those who are *outsiders* say. However, those who "linger" a little longer in the village begin to realize through the everyday landscape that there is something more[66] . This something more can be captured in the poetry of Elisa Lucinda when she says: "If I spend more than 24 hours in a village with love, I become a native, a grandmother, a midwife and a neighbor." Unfortunately, however, this something more, in the view of common sense, translates as Folklore[67] , or

---

including the interviewees' observations of the spaces; all idealization, interpretation and method is by Xavier (2008).

64 A Nietzschean allusion?

65 Bonnemaison (2002) cited above.

66 The writer, poet and actress Elisa Lucinda, from Vitória in the state of Espírito Santo, is a regular visitor to Vila and has her own home there. In her poetry we find, and at various times, clippings of the daily life of Vila, as experienced by her.

67 Folklore comes from the English folk - meaning people. For Ferreira (1999, p.922) 1- It is the set of traditions, knowledge or popular beliefs expressed in proverbs, tales and songs. However, this designation is extremely reductionist and presupposes a set of legends, myths, a framework without much contextualization, but always a disqualification. A dis-symmetry of values between urban culture and folk culture. Urban would be the developed, and folk would be the backward. On this, see Candido (1979, p. 21), as well as Garcia (2001)

something with less prestige, something bucolic. Even the communities themselves have embraced this *folk* discourse and today there is a "Folklore Association", with a representative elected with full honors. However, I think it goes beyond this reduction, as I've said elsewhere. In the interpretative sense of local culture, I understand that this adhesion was for reasons of participating in the power game and gaining political legitimacy among the "others" in order to act. This is very common in the game between unequals[68] . In fact, there is a legitimized institution, its roles, its sub-institutions and it represents a strong socializing connotation in the local context.

According to Silva (2006), social relations in small towns are marked by personality. These face-to-face relationships generally exert a very effective control over the members of a community. Clifford Geertz (1978), quoted by Silva (2006), treats culture from the perspective of a 'control mechanism', and argues that it "begins with the assumption that human thought is basically both social and public - that its natural environment is the family courtyard, the marketplace and the town square". Geertz (1978) also argues that it is necessary to pay attention to people's "behavior", to the "micropolitics" (exercised through Weberian social action) that inform Félix Guatari and Sueli Rolnik.

In line with Silva's thinking (op.cit.), it is possible to say that everyday semiotics, these microgeographies, are relationships woven through the network of reciprocity, and therefore affective, beyond relationships of simple exchanges, or purely economic exchanges. If face-to-face relationships have a social "glue", and these relationships weave a fabric that is permanently produced on a daily basis through and in micropolitics, then we can say that the landscape expresses this (semiotic) relationship. This landscape we're talking about goes beyond the contour, it's symbolic, alive and dynamic, and that's why it can also be a mark, a matrix, fixed and flowing.

According to Silva (2006, p.17):

"In the central square of a small town, "where everything happens" (in the view of its inhabitants), are the residences of the local elite, where they form "networks of prestige" with members of a certain social stratum who don't necessarily have economic power, but concentrate traditional families. This same place can serve as a place for young people to socialize on Sunday evenings, or even for religious festivals bringing together various income classes, united by religious belief. Thus, the appropriation of space is determined by the relationships established between its members, by the use of common symbols and codes."

With a closer look, and from a privileged vantage point such as the Praça da Vila de Itaùnas, for example, we can perceive the coming and going of people in their daily lives, in their spatiality/temporality. At every moment, new landscapes are formed and they express the *modus vivendi* of the society that builds its "place" there. In this sense, it is from this imagery that we can infer that the city has a "soul", an identity, as Tuan (19 83)[69] , Lynch (1999) and Silva (2006) tell us.

But who lends the "soul" to the Vila, or the city? Through his poetry, the poet Fernando Pessoa (1997, p. 49)

68 Cf. Xavier (2004) .

69 "What gives a place its identity and aura?" (TUAN, op.cit., p.4)

offers us a reflection:

"A butterfly flies past me. And for the first time in the universe I notice. That butterflies have neither color nor movement, just as flowers have neither perfume nor color. It's the butterfly's wings that have color. In the butterfly's movement, it is movement that moves. It is the perfume that has perfume in the perfume of the flower. The butterfly is just a butterfly. And the flower is just a flower."

The poet, in his sensitivity, tells us about the experience of the moment, in the same way that Tuan (1983) tells us about the experience of space in childhood. In fact, it is the experience of seeing that gives meaning to the given wavelength, which we identify as "color" and which gains meaning within a cultural system that calls something that flies, with that "butterfly" aspect. And in this system of meanings, a butterfly is something simple, light and beautiful. In this sense, the poet brings all the emotion to present a poetic position and exposition. Although color is decoded by sense perception (sight) and the process of cognition, in a given moment of space/time that we experience as humans, in a certain system of meanings, it has color. It's not the butterfly's wing, it's our experience that tells us all this. See what Lynch (1999) tells us about this experience of space.

For Lynch (op.cit., p. 4), humans structure space to identify their environment, and this is no different from other animals that move around. However, humans give space meaning. Their spatial experience goes far beyond the five senses. For the latter author, it is a system of meanings added to points of orientation, since each human brings a mental framework of orientation and a framework of significant images.

"This image is the product of both immediate sensation and the memory of past experiences, and its use is intended to interpret information and guide action. The need to recognize and standardize our environment is so crucial and has such deep roots in the past, that this image is of enormous practical and emotional importance to the individual. (...) a clear image allows us to move around more easily and quickly: to find a friend's house, a policeman or a haberdashery. However, an orderly environment can do more than that; it can serve as a vast reference system, an organizer of activity, belief or knowledge. "

In this sense, images - in the landscape - carry a charge of emotion that guides our actions, thoughts, feelings and is in our frame of reference for life.

On this Tuan (op.cit. p. 5) says

"we measure and map space and place, and acquire spatial laws and resource inventories through our efforts. These are important approaches, but they need to be complemented by experiential data that we can collect and interpret reliably, because we ourselves are human. We have the privilege of access to states of mind, thoughts and feelings. (...) People are complex beings. "

To illustrate, here is a testimony from a community member and her husband:

"On the feast of St. Benedict I went to the rehearsal... my daughter... but it was very good. If you watch the rehearsal... you'll like it.... Hail Mary! [with emphasis] it's so beautiful... I thought it was so beautiful... I cry... my daughter... If you saw it... and went to the rehearsal... you'd like it! They came to pick me up at two o'clock. I went to ...Anedir's house, my daughter Anedir. The rehearsal is there, on her farm... it's good to see, you needed to see it. Then she put the lights on and everything was aluminized... everything was beautiful!" [showing admiration and enchantment, joy]."

"Her husband complements: lots of people at the dress rehearsal! A fork to eat Oh! Hum, hum, hum,...Oh?" [showing joy and satisfaction, prosperity and abundance].

(Woman, 81, and her husband, 81 in January 2007).

You can see the satisfaction and joy, and how with each celebration this joy is revived and renewed. When we talk about the "feast", an image of colors, symbols, tastes and smells, all in movement, forms a whole and the landscape itself, perceived and experienced, is now an affective memory, a memory of "place", a memory of a space/time - which belongs to each person, but belongs to everyone's memory. In these terms, all the interlocutors, local actors, interviewed about the Ticumbi, Alardo or Reis de Boi games confirmed this landscape memory, brand and matrix.

The boy (16 years old), who gave us all the spatial and socio-cultural references of the village, brings with him this interesting experience, since he actively participates in all the social life of the village. Either directly or indirectly.

I: Do you play Ticumbi?

L: It's for fun! It's called a joke here in Itaùnas. They say it's... *jokes,* people say dancing, but the word dancing doesn't fit very well. It's more like play!

I: What does this game mean? You who are playing now, who are young and see the older ones, what do you think this means for the village?

L: That's right, that's a...*a tradition*, right, that goes back to the ancestors, that tells...that the *ticumbi is actually a story*. A story about two kings. Two African kings who...each wanted to celebrate Saint Binidito, right? Only one was pagan and the other was... The other was Christian, right? So they fought each other, and in the end it was the Christian king who won, who baptized the pagan king and the two of them took part in the feast together in the end!

E: And this story is told entirely in dance and music?

L: It's in music and also... the spoken part, you know. Spoken parts!

I: So it's not just music?

L: No.

I: Is most of it spoken?

L: So, the first part is music, music and dance. Then, in the middle of the game, we open the circle and they come in to recite his embassy verses. The Christian king is the king of Congo and his secretary. And the pagan king is the king of Bamba and he also has a secretary, right? He's the ambassador of each king, he's the secretary. Then, in the middle of this... dance, there is, right? Then, when they...declare peace, there's another part of dance and music.

E: Only by seeing, right?

L: Only by seeing it.

I: But do you know all the words? Of the story told?

L: We know more or less, because it's a lot, a lot of verse, a lot of verses.

I: You can't memorize everything?

L: It's hard to memorize everything, isn't it, because there's so much!

I: Is there someone who knows everything?

L: Usually the master, right, of the group knows everything.

I: That master you told me about, right? The new one?

L: YES.

I: Ah, then we'll have to talk to him.

L: He's actually already been a master of other groups, but it's because he's formed a new group.

E: The one in Santa Clara, right?

L: YES.

I: Then we have to talk to him.

L: He's Lelé's uncle.

(interview in February 2007, male, 16 years old. L: interviewee; E- interviewer)[emphasis added]

At another point, the interlocutor describes his vision of the "Reis de Bicho" folklore:

"L - at Reis de Boi we have the animals, right? The main thing the children like is the animals!

E- Who imitates animals?

L- Who imitates animals [agreeing], who has animal clothes and everything....

E- But I haven't seen you in animal clothing!

L- No, we were dressed as sailors. The Kings begin like this: outside the house, with the lights off. Everyone with the lights off and the door closed, the owner of the house. We begin to sing the praises that we call "door sound".

It starts at a very slow pace. It goes on to tell the stories of the Kings, that the Kings traveled, it tells the story of the birth of Jesus, of the annunciation. Very slow and long! It takes about 20 minutes or so. When the "sound of the door" finishes, the verse is sung for the master of the house to open the door and turn on the light. Then... as soon as the master of the house opens the door and turns on the light, there's the 'descante', which is a very fast and very cheerful march, right? It introduces. Then there's the march to enter the house and some other marches. Then, in the middle of the game, there's the presentation of the animals. It's the cowboy who comes to offer the animals to the owner of the house. That's why it's called Reis de Boi! [emphasis] First he presents the ox!

In our game, the ox dies and the cowboy gives it to the people. Always in rhyme... like the bacon belongs to Seu Caboquinho, the tongue belongs to Dona Domingas! He shares it out like this... Rhyming. Always in rhyme. He divides it up like this, rhyming. Then he charges off, circles the mill and introduces the other animals: animals like the she-wolf, the seahorse, the "two-faced" and many other animals!!! [his 6-year-old sister says: one at the back and one at the front? In the end he sells his wife, who is called Katirina. She dances with the men and charges them. It's really funny, isn't it? Nobody wants to dance with this woman. It's actually a man disguised as a woman! It's too funny! [laughing] Then she gets ready to go to another house.

This is how the Reis de Boi game works.

(Interview with male, 16 years old, February 2007, Vila de Itaùnas. L: interviewee; E: interviewer)

According to this interlocutor (16 years old, 2007), about the clothes he wore while playing the pandeiro: "I was de "marujo", the person who plays the pandeiro and sings is a "marujo". Now there are the animals' clothes. There's the animal's head and the animal's body, which is covered with a chintz cloth". Regarding the musical instruments of the Reis de Boi: "and there's the accordion player who sets the tone. Often there's a guitar as well and it's more beautiful! The only real instruments in Reis de Boi are the pandeiro and the accordion player." In addition, according to local custom, Reis de Boi is always performed at night and is invited to perform throughout the Conceiçao da Barra region and surrounding areas, because people like it so much. It is customary to offer a snack to those who perform Reis de Boi, but if the owner of the house has nothing to offer, Reis de Boi performs anyway, says Lucas.

In these interlocutors' speeches we can observe a "narrative of space/time", of spatialization in their expressions, because according to Certau and Mayol (1996, p.201) "through the stories of places, they become habitable. To inhabit is to narrativize. Fostering or restoring this narrativity is therefore also a task of restoration." Where the "event" is the one that is told, to paraphrase the aforementioned authors, the Vila has no history, it can only live if it preserves all its memories! I realize that we have to let the word go, the narrative about the *modus vivendi* presents the *place*. The fact is that "the *where* determines the *how* of Being, because Being means presence", in Heidegger's view (1992, p.90, apud. SANTOS, 2004, p. 93).

This colliding landscape - memory and matrix - of a time/place is also one of fixes and flows. In this author's interpretation, the flows are the processions, the procession of the Ticumbis, the Reis de Boi, and the fixed points are: Praça da Matriz, the Igreja Matriz itself, the pequi vinegar,

Fixed and flowing together interact and express geographic reality and so, in this joint interaction, they emerge as geographic objects that can be mapped. Taking this path of thought, I say that this conniving landscape, as well as containing the fixed and flows, has the attribute of being a mark and configures an imagetic matrix, since the image in the memory of the agents revives and consolidates the feeling of belonging to the place, we belong to something that belongs to us. Provisionally, we can understand that it is in this sense that the Vila has a "soul", and this soul is given through and by everyday relationships, forming a Landscape that is present, alive, because it has been lived, suffered and sung. If anyone doubts this, visit the ruins of old Itaúnas, or any other town that has been displaced (by a dam, for example). It's just a collection of artifacts, objects, an aesthetic landscape without a soul.

Finally, the religious festivals of Vila de Itaùnas cannot be seen as "folk", as a mythical tale, as a simple representation of a purely religious cult, plastered in Catholic doctrine. It is an institution of popular Catholicism, as we have already said, but it is also a political act, a cultural reaffirmation, an identity mark, in which there is a process of *reciprocity* as described by Mauss (1974), which we will talk about in more detail in another chapter. It is the expression of a worldview. It is the staging of a cosmogony, the hierophany of the world, as Eliade (1992) and Rosendahl (2001, 2008 a,b) show us. Chaos needs to be given order. The festivities of São Sebastiao and São Benedito, through their games, recreate and reinterpret the creation of a world that has meaning for the subjects/actors of the village. In these moments, they affirm and reaffirm tradition - as recurring social practices, and a way of dealing with time and space, inserting ritual marks in the form of particular activities or experiences. With the force of repetition, I say that these traditions, these doings with knowledge, are configured as a true *intangible heritage* (dances, choreography, rhythm, songs, melodies) and *material* (the indumentary white and colored clothing, the embroidery, the hat adorned as a mark, the colors, the flag, etc.) as Xavier (2004, p. 161) shows in his work on the representations of the communities of Saco do Mamanguà, Paraty, RJ.

5. The network of affections and other spatialities

The metaphor-image or even representation of the network has existed since ancient Greece, in various human societies in their mythological manifestations, as Musso (2003, p. 16) shows us. The image of the weaving of the fabric, of the mesh, of the interweaving of the threads forming the weft, inspires and represents the symbolism of interconnection, and was first used in mythology. The genealogy of net figures shows that this original reference persists to this day. The mesh of the net contains abstractions such as body-cosmos, nature and planet, society and organisms. Plato uses the figure of the king-king and compares the object of Politics to the fabric; at the turn of the 18th and 19th centuries, Diderot and Saint-Simom rethink modern politics using the figure of the network. Today, according to Musso (ibid.), from mythology to cyberculture, representations unite in an articulated configuration that lies between the images of the network technique, the metaphor of bodies and the figures of politics. Finally, what is considered here is that the "network" brings this image of the interweaving of threads, of the mixture, in an art of weaving with opposites (in a politics). The idea of capturing, but leaving alive, organizing and, above all, composing. In the 18th century, engineers and geographers formalized the search for the mathematization of the reticulum, thanks to the new tools that allowed them to geometrize space, a necessary path for the establishment of cartography; at the same time, biomedical practitioners use the same metaphor to understand the organization of bodies, of reticular structures (Marcello Malpighi's reticular bodies).

Musso (ibid.) also informs us that with the exploitation of tele-information techniques, such as the Internet phenomenon, the network has imposed itself as a totalizing (and totalitarian) figure that has evaded technological, political, scientific and other discourses. But he warns that we must not reduce it to a notion empty of all meaning or reduce it to a simple effect of fashion.

For geographer Claude Raffestin (1993, p. 145), the *image* or model is an instrument of power, or rather, every representation or construction of reality plays this role in all societies. This image, which can be a guide to action, has taken many forms and we have objectified it. We began to act more on the images, now *simulacra* of objects, than on the objects themselves that the image represented. In this sense, Raffestin (1993) urges us to rethink what systems of representation have to do with the play of power throughout history. For this author, cartography was born in the Renaissance, following in parallel with the birth of the Modern State, and quickly became an instrument of power and of Power, an important piece in the game. In fact, the geometrization of space carried out by the engineers and geographers of the 18th century was, as Musso (2003, p.26) puts it, the mathematization of the reticulum, thanks to the new instruments that enabled them to achieve this feat. This cartography privileged Euclidean syntax, i.e. the shapes and dimensions of mathematical beings, mobilizing three dimensions of space: the surface or plane, the line or straight line, and the point or moment of the plane. The combination of these elements results in images or representations of space. This structural game, in Raffestin's view (op.cit.), has satisfied the need for representation for a long time. But what does this representation of space have to do with networks?

Raffestin (ibid., p.146-147) gives us an interesting diagram of an actor representing his space:

Initially, the actor is situated at a point in space, a point from which they will represent space itself. The point is not, contrary to what it may seem, privileged in relation to the other elements - surface and line. It only provides the origin of the representation, that is, it provides the egocentric support for the representation, since the latter is always a manifestation of the 'I' in relation to the 'not-I', an explication of interiority in relation to exteriority. What do we have in this simplistic and yet sufficient scheme? Points that can represent the location of other actors or properties that interest A; lines that join other points and delimit a surface. (...) What is important to understand is the value of this scheme as a representation of a space for actor A.

This representation of space does not exhaust space and its dimensions, it is only the egocentric representation of A and does not take into account any actor other than A. It is the representation of lived space, the experienced, but egocentric. This representation communicates intention and material reality through a systemic system. Space is no longer space, but an image, a simulacrum of lived territory. Raffestin (op.cit, p. 147) says that the space that has become the territory of an actor is only such to the extent that it is taken up in a social relationship of communication.

In line with this thinking, if we look at this space not as static, but as a field of possibilities, actor A can draw up various types of weaves and articulate points with other actors such as B, C, D, etc., in networks; in this sense, Raffestin (ibid. p.148) gives us a scheme he drew up to illustrate the idea:

(...) from this original representation, the actor can decide to "build" various types of weaves and articulate all the points or just some of them in networks. He can decide to connect certain points, ensuring continuity between them by means of a system of junctions or, on the contrary, prevent certain points from being connected to each other by imagining a system of disjunctions.

What can we infer from Raffestin's demonstrations (ibid.)?

That the network we are referring to here is made up of joints of territory and social territorialities, spatialities, therefore actions and relationships that reveal power games, express a *modus vivendi*, an *ethos*. And whether we're talking about weaves, knots or networks, there are many explanations that lead us to understand territorial construction, the territorial system. It is well known that networks can also be technical, but they are still social. But in this chapter we will preferably talk about social networks, their junctions, their micropolitics woven through and in geography: from *géographicité* as an expression dear to the phenomenological geography of the Frenchman Eric Dardel (1899-1967), and expanded in other ways by Tuan (1983) and Berque (2000, 2004) and analyzed by Holzer (1992, 1997, 2001) among others. This notion is used to dimension the relationship between the human being and the Earth and the human being in its expression in a mode of existence and destiny. Imbued with this thought, I can infer that this geography of the social subject in a *state of community*[70] in the village of Itaûnas operates a spatiality that I call here a *modus vivendi* - an *ethos* - a differentiated society-nature relationship, a way of inserting oneself into the world, constructing an ever-dynamic landscape.

Taking a different approach, but still seeking this social meaning, Scherer-Warren (2007, p.29,30) shows us that the notion of social networks has strong roots in the social sciences. Among these, the author points to two main strands, namely: (i) the search in the notion of the network for an explanation of social structure,

70 Moscovici (1990, p.56) uses the term to express a feeling of community. This meaning coincides with Weber's (1973, p.141) "we call a social relationship a community when the attitude in social action - in the particular case, in the average term or in the pure type - is inspired by the subjective feeling (affective or traditional) of the participants in the constitution of a whole."

based on the theory of Radcliffe-Brown and his followers, from the 1940s onwards; (ii) the other found in this notion a way to describe the primary social relationships in everyday life, typifying these relationships as closed or open, strong or weak links, in line with the theories of Barnes and others. The first aims to explain the structure of the social (theory of explanation of reality), and in the second, what matters is the empirical observation of different forms or intensity of social relations in a given social field - kinship, friendship, neighborhood, religion, etc. What will interest us from these theories will be the second strand, where we will speculate on the forms of interaction of social relations in Vila de Itaûnas, as an inter-relational social field, in its multiple dimensions, including the spatiality of the agents in their expressions.

In this sense, Scherer-Warren (2007, p.40-41) tells us that the forms of sociability in networks, as well as the respective relationships of identification or asymmetries of power, can be named in different ways, such as: reciprocity, solidarity, strategy and cognition. In order to demonstrate this, the latter author presents the work of Vargas (2003, p.8, apud. SCHERER-WARREN, 2007, p.40) on the neighborhood network in a poor neighborhood in Santo Domingo; "networks have a reciprocity meaning to the extent that activities are exchanged, services and favors are distributed." For the latter author, the analysis and notion of social networks based on the category of reciprocity, which I will use here, has been especially useful in studies aimed at the social relations of local daily life, this daily life demonstrated by Vargas (op.cit., ibid.).

This typification of the network is based on a characterization of the activities that generate networks or are carried out in networks, through the network, and examines how these different networks connect with each other, generating a total network of networks, which are: networks related to the life cycle, survival networks, networks of extension and support for domestic tasks, networks of leisure treatment and affective support, the presence of networks in migratory flows. In these types of networks, power relations are tacit, they are not explicit, and being in the network, this hierarchization is accepted as a social normality (see the Ticumbi Institution for its hierarchies - its social roles). However, this hierarchization, says Scherer-Warren (2007), can turn into resistance to external interventions in their daily lives, even if they propose to eradicate local poverty, for example. The author shows that even if the state takes positive action to eradicate poverty, with public management incentives, if they don't take into account the presence of networks that resize aid and interventions, the return effects will not be as expected, due to the strength and social power of networks within a community - social capital[71] (BOURDIEU, 1991, p.149-150).

In this line of thought, we are going to present the social networks in the village of Itaùnas as Networks of Affection, since the connections and processes of the constitution of this network, in my ethnographic reading, are based on ties of affection and reciprocity, as described in Marcel Mauss' Gift (1974): giving, receiving and giving back; as also demonstrated in the works of Vargas (ibid.) through Scherer-Warren (ibid.).

But what is a gift? For Mauss (ibid.) and his followers, this gift is taken as a *total social fact*, in other words: a fact that permeates all social relations, and not just the economic area, as modern capitalism supposed, since the exchanges in the gift are not simple exchanges, since the things exchanged have a soul and therefore have

71 "Symbolic capital is a cognitive-based capital, based on knowledge and recognition," in short, according to Bourdieu (1991, p. 150)

to return to that first donor (gift and counter-gift) (GODELIER, 2001).

Marcel Mauss (1974), when analyzing Melanesian and North American Indian societies[72] , realized that in these so-called "primitive" societies, everything was mixed together, everything constituted life itself, and for this reason he called them "total social phenomena"[73] . In this author's view, the *religious, legal and moral institutions of* these societies were interconnected by the process of the Gift, since institutions are *political and familial at the same time as economic.* For Mauss (ibid., p. 44, 45) it's not pure individuals who exchange, but moral people such as clans, tribes, families - who confront and oppose each other, either in groups, face to face, or through their chiefs, representatives, or both at the same time. Well, what is exchanged? Mauss (ibid.) tells us that not only goods and wealth, movable and immovable and/or economically useful, are exchanged. But it is mainly exchange:

"**Kindnesses, banquets, rites, military services, women, children, dances, festivals, fairs** in which the market is only one of the moments and where the circulation of wealth is only one term of a much more general and much more permanent contract. In short, these payments and counter-payments are made mainly voluntarily, by gifts and treats, although they are, in essence, strictly obligatory, on pain of private or public war." (MAUSS, 1974, p. 45)[emphasis added]

This system is referred to by the author as a *system of total services*, where these services and counter-services (reciprocities) are done voluntarily, for pleasure, for gifts, although they are in fact obligatory, under penalty of private or public war, since a tacit social rule has been broken. Godelier (ibid.), in his studies on the work of Marcel Mauss, argues that there is a "movement" of things and that they don't move on their own; in fact, it is the people in this network of gifts and counter-gifts who, believing in the magic or soul of the thing given, make it circulate. Since "giving seems to simultaneously institute a dual relationship between the giver and the receiver. A relationship of solidarity" (MAUSS, ibid., p. 23). An obligation is established and the network is formed. Let's see how it happens in the case study of this work.

In the village of Itaùnas, the festival to commemorate the saints - Saint Sebastian and Saint Benedict, the apex of Ticumbi - a hierophany[74] , brings this dense social network - the network of affections - into its process of institutionalization and implementation. In the understanding of this work, the social network constructed operates through its agents, constituting a support for the organization of the Ticumbi Institution and the sub-institution of the healing arts, in an action that makes the festivities effective, and presents a complex process of high intersubjectivity.

But how does it operate? In the context of this work, this network operates through the category of *reciprocity*, as in the studies of Mauss (1974) and his successors, which we have already discussed. This category is situated within what Scherer-Warren (2005, 2007) calls *social movement networks*, which are characterized by the complexity of the connections that operate in a symbolic and mainly solidary way. These networks engender an identity and cultural process in a dialogical way.

72 Mauss (ibid.) analyzed the work of anthropologists Franz Boas (1897) and Malinowiski (1922).
73 This is a legacy of Dürkheim's social theories of the sacred, which influenced Marcel Mauss, and with which Lévi Strauss disagrees (see discussion in SIGAUD, 1999 and GODELIER, 2001).
74 Manifestation of the sacred, in a sacred space (ELIADE, 2001).

Sociologist Scherer-Warren (2005, p.78-79), whose work on social networks is well-established, clarifies the difference between networked collectives and networks in social movements:

"The category "collective in network" refers to the connections - initially communicational and instrumentalized through technical networks - of various actors or organizations that want to disseminate information, seek solidarity support or even establish joint action strategies such as, for example, the *links* and connections that NGOs promote among themselves or with other relevant political actors, through the internet or other alternative media. This wider network of social movements. For example, the *online sites of* feminist NGOs, gender discussion lists, virtual or face-to-face women's forums, feminist reflection groups, women's civil associations, etc, all of which connect feminist activists or sympathizers, are nodes (of the network - a network is a system of interconnected nodes) or, in other words, networked collectives of the feminist movement, which is ultimately a network of networks of identity collectives.

The "networks of social movements", on the other hand, are complex social networks that transcend empirically delimited organizations and that connect individual subjects and collective actors in a symbolic, solidary or strategic way, whose identities are constituted in a dialogical process: a) of social, ethical, cultural and/or political-ideological identifications, that is, they form the movement's *identity*; b) of exchanges, negotiations, definitions of fields of conflict and resistance to adversaries and the mechanisms of discrimination, domination or exclusion from the limits of this systemic situation in the direction of the realization of alternative proposals or projects, that is, they set their *objectives*, or build a *project* for the movement. "

It should be clear that these two networks are not mutually exclusive and do not exist in isolation, since both are idealizations for sociological understanding, and can perfectly coexist in solidarity. But according to Scherer-Warren (ibid.), in order to better characterize the collectives in social movement networks, it is necessary to look at the historicity of their formation to find the dialogical dimensions between the collectives in action, in the sense of actions and the emergence of new collective intersubjectivities in movement networks. It is necessary to understand how individuals become subjects of their personal destinies, and how they become political actors through connections in networks. The aforementioned author emphasizes that the actors and the respective social movements to which they belong present themselves as forms of resistance and propositions in relation to the oppressive cultural codes, such as the informational and other codes that govern their lives. This approach considers the relationship between subjects and collective actors and their transformation into social movements, based on a triple dimension of information system (IS) networks: social, spatial and temporal.

In the case of the networks of affection[75] in Vila de Itaùnas, we need to understand it as a social movement network, an identity network, since it is a religious action strategy that is also political (different from party politics). This social movement gives rise to and maintains a[75] Institution[76] - the Institution of Ticumbi, and its sub-institutions that affirm and reaffirm values and maintain the[77] Afro-Brazilian tradition of the Itaùnas,

75 Peirano (2003) presents us with the rituals of the kula, a ceremonial exchange ritual (complex system), as a social network invested with morality and affectivity.

76 A social institution, according to Berger and Luckmann (op.cit.), has its origins in habit, since any frequently repeated action becomes molded into a pattern that can be reproduced with economy of effort. Habit also implies that the action can be performed in the future, and in the same way. These processes of habit formation precede all institutionalization. Institutionalization occurs whenever there is a reciprocal typification, and these exercised in this reciprocity give the typical character not only of the actions, but also of the actors in the institutions (social role), since these typifications are *shared* with others and are accessible to all members of the particular social group in question - in this case the Ticumbi of Vila de Itaùnas as a material and immaterial heritage.

77 In the sense of Hobsbawn (1997, p.9), traditions are invented and are understood as "a set of practices, usually regulated

their roots, their origins.

This tradition is thought of here as a way of dealing with space/time, inserting ritual marks in the form of particular activities or experiences. These ritual marks give continuity to the past, present and future, structured by recurring social practices. These traditions are configured as intangible and material heritage knowledge/doing, an important social capital for future generations (XAVIER, 2004). This rural society has yet another space/time relationship - *slow time* - a place of space, home to the gods, a place of enchantment and sacredness. To understand this, you have to linger in this *place*.

In this book's understanding, this happens through the connections that establish the nodes and build their spatialities through and in the relationships of the subjects in the network of affection. The nodes can be represented by the *festival-goers*. The festival-goers are those who bear the economic costs of the festivities, whether by donating animals for raffles, auctions, or even foodstuffs for the participants' meals. The festeiro is a fundamental part of the institution, of this festival, since he is the one who organizes it, pays for it, receives it and maintains a space for the congos to rehearse. A congo group always makes a special obeisance, a song of thanks at the festeiro's house. Spatiality is expressed by all these connections and networks of affections that effectively make Ticumbi possible.

For example, the route taken by the Ticumbi do Bongado congo group (the oldest in the village of Itaûnas), whose celebrant is Dona Maria Catarina, necessarily passes by her house, bows; she offers a treat, a snack, a drink and they reciprocate with songs of thanks. This happens either on the day of the Sâo Benedito and Sâo Sebastiâo festivities, or at another performance that the group does in the village, or even at the rehearsal site that the festival-goer keeps on the Agua Branca estate, which she owns. However, it should be made clear that these are the festival-goers, but there are also the festival-workers who have less social visibility, but who also do their bit, donate their labor, and contribute to the festival, to the rehearsal, each with what they can. Each Ticumbi has an itinerary (it's a thread), one or more festival-goers (nodes) and the whole of the Ticumbi is the very network of affection that constitutes solidarity expressed spatially, i.e. in its spatiality.

We always interpret this donation, in general, as a spontaneous donation, but one that demonstrates as a social action (in a Weberian sense) a strong political tactic, which gives great visibility, an insertion strategy that makes people who are not from the Vila, become considered to be from the place: a cohesive part of the space of relationships in movement (the game of established and outsiders described by ELIAS, 2000). This strategy was observed during my fieldwork, in which I noticed that some social subjects from outside the village (outsiders), from other cities and states, and now living in the village, used this social action tactic in order to be recognized and establish themselves as part of the *place* by participating in Ticumbi in some way.

This tactic and political strategy has worked very well, since there has in fact been a local acceptance of these people who now act and move around the village like the *established* ones, even though they have a strong *outsider* accent and significant educational and cultural differences. There are other cases of *outsiders* who

---

by tacit or openly accepted rules; such practices, of a ritual or symbolic nature, aim to inculcate certain values and norms of behavior through repetition, which automatically implies continuity in relation to the past. "

have frequented the Vila for many years, in general they are from Vitória, but have no fixed residence in the Vila, but have practically been adopted by a family. In this sense, they end up joining the network of affections, as they are part of the institution as an official collaborator, and every year they make their contribution in the form of kindnesses such as: transporting people to the festival, arranging a medical admission in Vitória for a patient from the village, an acquaintance, an adopted family member, doing small favors as a counter-provision of the network of affections linked to Ticumbi for a relationship beyond the village of Itaûnas.

During participant observation, I met these social subjects and realized that in many cases they shared a great deal of affection with the family that had "adopted" them (the reference comes from the social actor interviewed). In this sense, they said that they felt completely "at home", as in "their place", even though they admitted that it wasn't, and showed strong cultural differences with the members of the Vila. In my opinion, this is a very interesting picture to observe, given that some of these *outsiders* were more resistant to the first contact[78] (act of avoidance, as in ELIAS, 2000), than the *established* ones. It was in this clash, on the edge of this tension, that I realized the strong strategy of power and manipulation that some outsiders camouflage with actions seen as positive by the established. Meanwhile, I witnessed tensions and conflicts between groups of *outsiders*, as well as between these established *outsiders* and the natives[79] , showing that peace in the Vila is always a negotiation, interspersed with silent conflicts, but which strongly interfere in everyday relationships.

In relation to the donations made by festival-goers, Zaluar (1983, p.13-14) presents a study on popular Catholicism, in which she argues about the category of festival-goer and their place in the game of positions within rural communities. This author understands *popular Catholicism* as part of the Brazilian rural milieu, and its devotees do not conceive of it as a pure theory, disconnected from earthly things; popular Catholicism is a religion focused on life here on earth, and for this reason it is a practical religion. The religious practice in focus takes place mainly through the execution of promises and the holding of festivities in praise of the saints, as a language that expresses in a space/time the ideas and motivations of the agents regarding the opposition and complementarities existing between people and between the various social groups. Zaluar (1983, p.15-16) noticed in his work that the[80] communities were abandoning the more traditional patterns, the customary rules of the saint's feast which is represented in the figure of the festeiro - the one who is the guardian, the one who pays for the feast, and organizes it as a consideration for a moral debt to the saint. Instead of donations, bingos and auctions to pay for food on the days of the festival, the church started organizing barriquinhas selling "food and drink". In addition, the dances, revelry and forrós so criticized by the priests, who saw them as a desecration of the sacred event, are now becoming scarce in the contrasts between urbanization and the countryside, which brings with it other values. Even the processions, the promises always made at traditional festivals, seemed to be losing their character. In the new urban context, the ex-votos, the pilgrimages to the sanctuary towns[81] gave way to saints' festivals. In this sense, the sense of vicinal cooperation for certain tasks

78 For the outsiders who acted as established, I as an observer was a 'pure' outsider and the act of avoidance, including looking, speaking, was widely used.
79 In the language of outsiders.
80 In a Weberian sense of belonging.
81 On these sanctuary cities, see Rosendahl (2001, 2008 a, b) and the notion of hieropolis.

- such as *mutirão* - also disappeared.

However, in the rural society of Vila de Itaûnas we find all these forms of cooperation and religiosity. Even with two *lanhouses* installed, with foreigners opening their *nightclubs*, with the Germans, French and Italians trying to dance forró and buying up all the available land, and pushing the locals further and further away from the village square - where the parish church stands, and the trunk of the pequi vinegar tree - geosymbols - the privileged setting for festivities and celebrations (religious and political). There is a cultural resilience that impresses.

The donation (be it anything), or counter-donation, that the festival-goer makes, has in the understanding of this work the reciprocity - of the Gift. The donation is for the saint (a mediator between God and man), be it Saint Sebastian and/or Saint Benedict, expecting a gift, a blessing from the patron saints - a divine blessing. This blessing and gift can come in the form of good health for the festival-goers and their families, a good harvest, a good catch during the year, a successful year for their business, etc. Over the years, what is asked of the saint has changed rapidly, due to changes in work activities and radical needs. If devotees used to ask for more fish and a better harvest in return, today, with increased commercial activity, the request continues in this direction (Zaluar, op.cit).

The activity of festival-goers donating to the Dom is very common throughout Brazil, especially where this religious segment of so-called popular Catholicism exists, and has already been reported by authors such as Brandao (1978, 1985), Zaluar (1983) Couto (2003), Calàbria and Silva (2008), Gabarra (2007) and others. For Gabarra (2007), the reign of the Congo is a type of religious association that proliferated in the Minas region, as well as in other parts of Brazil, in the 18th and 19th centuries, and its objectives go beyond the religious and folkloric/festive, and are also political. On the other hand, the geographer Corrêa (2008, p. 249) also presents an interesting work on the political aspects of the institution of the Brotherhood of the Good Death in Cachoeira (Recôncavo Baiano), which would appear to be solely religious in nature. In the case of this book, the political side of the congo (also shown in Dom - system of total services) was very evident in the village of Itaúnas, confirming this last piece of information.

We can see that the congo is a form of reign without sovereignty, where there is a king, subjects, a court, and it presents itself as a form of dialogue between the institution and the local powers. The members of the congo claim to be descendants of Africans who were slaves in Brazil, subjects without recognition, without a nation. The congada, or the Congo, or in the specific case of Vila de Itaûnas, the Ticumbi - a gathering of congos, are perhaps a way of organizing themselves to govern themselves, denoting a form of resistance that has been perpetuated over time, with re-readings of their past. Reformulations of being in the world and a special way of spatializing the world, a *modus vivendi*, which is expressed in the Vila through the signs of colorful clothing, a hat adorned with ribbons and many flowers, the drum, the tambourine, the viola, their beliefs, their songs, the sieves decorated with ribbons and multicolored chintz cloth, the animals of the Reis de Boi, the dance and turns of the women in the jongo, the magic words of the prayers and blessings, the magic baths with the fragrant plants[82] , etc. Those in the Ticumbi Institution are united by a collective identity, a collective will, and they move and expand their actions through this networked social circle that they build through and in everyday life. Whether through their actions as Congo, or through the sub-institution of the healing specialists.

In Vila, during the major festivities in honor of the saints - Sao Benedito and Sao Sebastiao - various other congos are invited to take part: Congos from Conceiçao da Barra, Congos from Sao Mateus, Congos from Serra and Congos from Vitória, as well as Reis de Boi groups from Conceiçao da Barra and women's Jongos groups from the surrounding area. The networked social movement makes these connections through the festeiros and festeiros workers. In this sense, information circulates, words circulate, ties are joined in a process of solidarity with the same goal, so that all the congueiros who go to the Vila receive food, drink and accommodation for the three days of the event. They receive this support via festeiros and festeiros obreiros, who mobilize a great deal to make the event happen. It's worth pointing out that, in general, the people who take part in the congo are simple people who can't afford to pay for it. Often, the municipalities contribute very little, and the church nothing at all, since part of the festivities are considered profane. The Vila de Itaûnas district doesn't have a priest for its parish, according to informants, and the priest from Conceiçao da Barra only comes to say mass and take part in the procession of Sao Sebastiao. In the afternoon he leaves and doesn't attend any other festivities in the village, since the festivities take place on three consecutive days. In this case, we can see the disregard and lack of interest in this movement of faith.

It takes a huge mobilization of solidarity and a real network of affections for the festival to take place. Some traders from other towns, other festival-goers, out of respect for the saints and their faith, who participate mainly for the sake of the gift, contribute by renting buses to transport the congeiers, as well as snacks and drinks for the trip. The local festival-goers provide all the accommodation (food, lodging, chemical toilets, drinks, transport) for everyone on the days of the festival, but also on the days of rehearsals, and at other festivities where the Ticumbi congos perform, as well as donating animals for the raffle, which helps economically with the upkeep and costs of the festive ritual. In this book, the donation made by festival-goers (major and minor) is interpreted as a political act of gift-giving - the Dom, as configured by Mauss (ibid.) and expanded by his current successors such as Godbout (1999), Godelier (2001), Caillé (2002), Martins (2002), among others.

At the time of the ritual celebration, these people feel in communion and present themselves in the interpretation of this work as tireless, with a sacred feeling bringing them together. Ritual, according to Turner (1974), is a practice that follows certain procedures and takes place in moments marked by a break in routine. The Ticumbi festival, or the ritual consecration of Sao Sebastiao and Sao Benedito are seen here as ritual processes.

This feeling of *communion* within rituals was described by Turner (1974, p.116) as *communitas*. This author theorizes that in a liminal process (of passage), or liminal phenomena, societies that were previously organized in terms of caste, class, or hierarchical orders, or segmentary oppositions, in these special ritualistic moments

82 We'll take a closer look at these plants, magic words and blessings in the next chapter.

fragment these ties, these terms, and begin to act within an area of common life. It is a "moment situated in and out of time", in and out of the profane social structure, and reveals, in a tacit and ephemeral way, another social bond, a communion as an unstructured or rudimentarily structured *comitatus* - a community, a communion of equal individuals who jointly submit to the general authority of the ritual elders (TURNER, op.cit., p.119). In other words, this feeling operates a different kind of social relationship, where there is a common area of life, where everyone helps each other at the same time to fulfill a common goal. In this process, the social positions of more or less can be inverted, as they transgress the rules of the social and are accompanied by an experience of unparalleled power. In this *communitas* moment, the weak are endowed with ritual powers to the point where the subordinate becomes predominant. And it is in this case that a fisherman from a simple life, or a small farmer from the village of Itaûnas becomes the King of Congo, and/or a King of Bamba, with all the investiture of the position. At this ritual moment in social life, the party merchant is nothing more than the King of the Congo, since he serves the King, invested with the representation of the sacred. The passage from a lower situation to a higher one is made through a limbo of absence of status, and in this process, according to Turner (oc.cit.) the opposites constitute each other and are mutually indispensable.

Men, women and children dance and sing with great joy, sharing this *communitas* ritual moment, while at the same time working together to make the ritual happen, since everyone benefits from it. We can see that they gain in the depth of their relationships, in the strengthening of the social bonds of belonging to a group, to a geographical place, but also a social place. Remembering, as Turner said (op.cit., p. 119): "Iiminarity implies that the high could not be high without the low existing, and those who are high must experience what it means to be low". Furthermore, by donating their free time to the festivities, they are giving to the saints and hoping for a blessing from the patron saint, St. Sebastian, and the saint of devotion, St. Benedict. The saints seem to act as mediators between God and man and in this case the sacredness of the donation is legitimized in the festivities where the sacred and the profane come together to express a way of being in the world, a *modus vivendi.* The families of the village take an active part in the ritual festivities, and I interpret this participation as the category of *conatus* (BOURDIEU, 1997, p.176), a reproduction strategy, a drive of the family, of the house, to perpetuate itself. It is also a strategy with a strong identity. See two typical Itaunense families in figures 32 and 33 below.

Figure 36: family participating in the Ticumbi Institution. Photo Xavier (2007).

Figure 37: family taking part in the festivities and folklore of the Ticumbi Institution, a gathering of faith. Photo Xavier (2008).

Figure 38: family participating in the Ticumbi Institution. Photo Xavier (2007)

5.1 The Network of Affection: an ethnographic perception of the field.

Berque (2004) tells us, as we saw in the previous chapter, that this landscape is both a mark and a matrix, reaffirming the cultural ties made dynamic through discourse, through actions and through the networks of affection woven into everyday life, transforming a banal space into a *place*. In the village of Itaûnas, this image of a *place* remains in the affective memory of the agents as a *mental landscape*, a *landscape of mark and matrix,* a place of reference for ancestors, life and personal destiny, which can be shared as a collective destiny.

In my fieldwork experience, I found this expression not only in the speeches of the older informants, but also in the speeches of the younger ones who "had to leave"[83] to go and study in Vitória, looking for the Federal University of Espirito Santo, and returned to the village after finishing their studies. This is the case of a young woman, 24 years old, single, the daughter of a family from Itaú, whose father is an artisanal fisherman, 53 years old, and whose mother is a housewife, 51 years old. This young woman completed her university studies in Vitoria for four years and returned to the village. In this interview, she says that the image of the village, her friends and her family was a constant in her mind, and that's why she decided to come back. She's dating a guy from the village and they intend to get married and continue living in the village, even if she has to work in another city, as she already does. For this young woman, as for others with whom I have had contact during participant observation, the village of Itaùnas represents her place of destiny. The young woman's whole family, like other young people in the village, takes part in the Ticumbi festivities, including helping to distribute and prepare lunch and dinner, the hot meals that are served to all the guests and to the organizers of the festival and controlled by a password. This donation of free time can be understood as the Gift, and this family is considered in this work to be what I have called a *worker-festeiro,* since they don't participate by donating animals for the raffle, nor with transportation, nor with the cost of food, but they do participate by donating their free time to make the festival itself a reality.

Another interpretation of this Network of Affections was perceived in artisanal fishing through the *piem.* Fishing in the village of Itaûnas is a traditional practice, in fact a craft that preserves a *body of* knowledge passed down through the generations by word of mouth and by practice, like the vast majority of master crafts that we find, for example, in societies with artisanal crafts in Brazil, as well as Portugal, Italy, Peru, Chile, among others.

The fishermen in the village are linked to the Z01 colony in Conceiçao da Barra, ES. According to informants, there are around 140 to 200 fishermen with "fishing licenses", but around 40 "active" fishermen. The most commonly caught fish are hake, sea bass, sea bass (summer), mackerel and puffer fish (winter), depending on the season. According to the fishermen, artisanal fishing has suffered a lot in the village because of the trawlers, large motorized boats that do balloon fishing (shrimp balloons) and destroy the fishermen's nets, as well as destroying their best "fishing grounds" - those places that are well guarded by the fisherman and where some fish hide at a given time. Often the fish are 'trapped' as the tide goes out, and that's where the fisherman gets a

83 In the local expression.

good catch. These boats have caused enormous damage to local artisanal fishermen, discouraging the younger ones, as many fish have disappeared, such as peroa. An account of the interview:

"The lack of fish is a general complaint, there was a case of a retired fisherman who said that in the past, about 20 years ago, he caught 970 turkeys in one morning. Nowadays, if they manage to catch two or three turkeys, they consider that they have made a good catch, due to the scarcity of fish" (fragment of an informal conversation).

"And then there's the trawling. They're wiping out the rest of the fish. The boats come when the fishing is closed here in Espírito Santo, and the ones from Bahia come here and spend three months trapping fish. When the fishing closes, the fishing opens in Bahia, the ones from here close, they come from there to here, when that's it, they keep on dragging all year round, they don't stop, do they? There's no peace! But apart from that..." (Fisherman, 45 years old, interviewed in 2007)

When asked about monitoring, he replied:

"There's nothing, miss. It's really isolated here, it's the park every day, it's just that it's a pain in the ass, even though IBAMA doesn't have one, which we've asked for a coastguard... we always ask for one so that we can see who's who, many fishermen say they're fishermen and they're not, and others say that these boats were closed at the time of the defense, nobody respects them, there's no inspection, there's nothing. It's the fishermen themselves who have to do the monitoring. If we talk to them, it's nothing! And they scrape the nets off us, in the sea, and last year I lost eight pieces of net." (Fisherman, 45 years old, interviewed in 2007)

When asked *where* he fishes, *how* he fishes and *who he* fishes *with,* he replied:

P-"I leave the house at four in the morning. By the time I start putting up the net, it's half past five, six in the summer!

And you leave it until what time?

One hour, two hours!

I-Are you going to pull?

P- I'm going to pull.

And who's going with you?

P-Myself, my friend Dario and Caca, my nephew, always go.

Are you always a friend? A companion?

P- Always a friend, a companion.

And do you have someone to help you pull, or just you?]

Just us. On the edge when we get to the edge there on the rocks there... in ***Lavadia***. **There are some companions there to help, but they don't always go...** We butemo and they even nicknamed us ***piem!*** **Then they go over there and help us gather the net and we give them some fish.**

I-What do you call it over there? Lavadio?

Lavadia!

E- Why?

It's because that's where the water comes, hits and washes away.

Is it coral or rock?

P- It's rock.

E-Ali isn't coral?

Q- There on those rocks? It's a rock!

E-You fish from a waiting net, don't you?

Waiting network, yeah!

E- Then there's the line fishing too?

Line fishing, isn't it? It's fishing that we do too, sometimes, when it's allowed too... Now ***gruzera fishing*** is forbidden, right?

E- What's this gruzera fishing like?

Gruzera are hooks that we put on the rope like this... stretched out like this, then every third arm you put a hook and put about ten, fifteen hooks on the rope like this, you put a big stone at the end and another one at the bottom. Then the net is stretched out. It's to catch grouper, they hunt them, but it's really rare that they hunt them these days, there aren't any on the beach now, there aren't!!!" (Fisherman, 45 years old, interviewed in 2007; P- Fisherman, E- Interviewer). [emphasis added]

You can see that he names the form of collaboration in pulling in the net as *Piem* - in fact an act of reciprocity. He also identifies the stone, the rock on which they arrive by boat as *Lavadia* - a geosymbol that identifies a place - a topos.

From the point of view of the interviewees, and from what we were able to observe during the ethnography, the piem is a help, a collaboration that is given or provided to an uncle, a friend or a brother of the fisherman when the boat arrives, the net is collected and the fish are sorted. See figure

Figure 39: representation of the *piem* helping to pull the canoe, Praia da Vila de Itaûnas. Photo: Xavier (2007)

Figure 40: The nieces helping the fisherman *with the piem.* Photo: Xavier (2007)

This collaboration is not considered work and is done spontaneously, regardless of gender, whether it is a friend, relative or compadre. From this point of view, it can be seen that in general there is a relationship of affection in the piem, which brings the subjects together through a practice, an action, and concretizes a relationship of solidarity, since in the piem the subject offers himself in a relationship where solidarity (from the one receiving the help) and reciprocity (from the one giving the help) are expected, typical of the threefold Gift: giving, receiving and giving back. Retribution takes place when the fish have been sorted, the net collected and the fishing gear put away; at this point the boat owner distributes some of the fish to those who took part in the piem, so that it also functions as a redistribution of goods, as in the Gift. A gift is a voluntary act, says Godelier (ibid., p.23), individual or collective, which may or may not have been requested by the person receiving it. In our Western culture, unsolicited gifts are valued, but this attitude is not universal, says the author[84] . Look at the dual relationship that is established:

> *Giving* seems to simultaneously establish a dual relationship between the giver and the receiver. A relationship of *Solidarity*, because the giver shares what he has, perhaps what he is, with the one to whom he gives, and a relationship of *Superiority*, because the one who receives the gift and accepts it is indebted to the giver. Through this debt, he is obliged and therefore to a certain extent dependent on him, at least until he is able to "give back" what he has been given. Giving thus seems to establish a difference and an inequality *of status* between donor and donee, an inequality that in certain circumstances can turn into a hierarchy: if this already existed between them before the gift, it would express it and legitimize it at the same time. Two opposing movements would therefore be contained in one and the same act. The gift *brings* the protagonists *closer together* because it is a sharing, and it *distances* them socially because it turns one of them into the other's debtor. You can see the formidable field of maneuvers and possible strategies virtually contained in the practice of the gift and the range of opposing interests it can serve. The gift is, in its very essence, an ambivalent practice that unites or can unite opposing passions and forces. It can be, at the same time or successively, an act of generosity or an act of violence, but in this case violence disguised as a disinterested gesture, since it is exercised through and in the form of sharing.

In this sense, there is no begging in the village, and no one goes without food, since most of the year there is fishing or shellfish gathering, so one neighbor always helps another in an act of solidarity and reciprocity.

84 Take the case of our Western society, even yesterday, but still today, where a man who wants to marry a young woman must ask for her hand in marriage, and ask officially through a family ritual (Cf. GODELIER, ibid.,p.23)

With a lot or a little fish caught, it is always distributed and (re)distributed to a neighbor, a relative, a compadre, a friend. When these people come together to work collectively, they form a network of solidarity, since they form strong social bonds. It seems that in the gift, the thing given has a "soul" that returns to its giver, or needs to return. Godelier (2001) informs us that there is a fourth obligation in the Gift: the gifts of men to the gods and to the men who represent the gods. Take the example in Itaùnas: the fisherman receives a gift from the gods, God or the saints, which is fish on a good day. He in turn, in an action of contradiction, gives it to those who helped him to collect the boat, gather the net and sort the fish. And those who help him in the solidarity act of piem expect retribution, which at first will be the fish, but it's much more than that, since they receive and donate something sacralized, which has a soul, because they received it from the gods, from God, from the saints. They believe that they have received the divine gift, and in turn they will redistribute the fish, making the gift circulate. The Gift works to circulate the thing - the thing that has a soul. In fact, the soul of the thing given (the thing being the object, the artifact, the fish, the magic word, the free time dedicated to mutirao work, the piem, etc.), is an imaginary reality, according to Godelier (ibid., p.104):

Its content is ideas and symbols that give the object a social force, a force used by individuals and groups to act on each other, either to establish new social relationships or to reproduce old ones. The imaginary, immaterial content of the things given is in no way reduced to the simple presence of the donor in the thing given. It is, of course, because the things given are "never completely detached" from their owner, that they carry with them something of his being, that through them people connect, commit themselves. They are "personal" relationships that are established, people who commit themselves. And the thing given is a guarantee of their commitment.

Still for Godelier (ibid., p.161), the gift becomes a sacrifice to the spirits and to the gods, within what Mauss called the *fourth* founding *obligation of* the gift practice. Belief in the soul of things amplifies, but also magnifies people and social relationships, since it sacralizes them. For if things have souls, it is because divinities, gods and saints, who are usually invisible, live in them, operate through them and circulate with them among people, sometimes connecting to some, sometimes to others, but always connecting them to themselves. This belief in the soul of things not only broadens, but substantially alters the nature of relationships and their meaning. It metamorphoses them, says Godelier (ibid., p.161). In this author's view, humans, who were once actors, are now presented as being acted upon by the objects they give or receive, subject to their wills and their movements. "It is no longer (only) human beings who act on each other, with each other, through things; it is things, and the spirits that animate them, that now act on themselves, through humans" (GODELIER, ibid., p. 162).

Perceiving these exchanges as more than just exchanges, but as a moral attitude towards life in relation to others, but not just to others, but to the world (human and non-human), a type of society-nature relationship, a *modus vivendi*.

The cosmos becomes the anthropomorphic extension of men and their societies. The individual is linked to the entire universe, which surpasses him and which also contains and surpasses his society. But at the same time, and inversely, the individual contains within himself, in a certain way, the whole of society and the cosmos. The microcosm of the individual contains the macrocosm that surrounds it and is, at the same time, contained within it. The part is the Whole, the Whole is whole in each of its parts. Each individual and the cosmos is like a mirror of the other and every action on one must act on the other. The whole world, including men, has become "enchanted" (GODELIER, 2001, p.160-161),

This *modus vivendi* can be seen in the expressions of local spatiality, such as the Reis de Boi festival, where the animals (the snake, the dog, the ox, etc.) talk and play with the men, scaring the children who run around, at once happy and fearful. The Ticumbi also performs at the congos meeting, in honor of Saint Sebastian, the patron saint, always accompanied by Saint Benedict, the saint of the "blacks", the saint of the poor. Flowers and colorful ribbons appear all over the congueiro's clothing, whether in the multicolored headdresses or in the decorations that adorn the route along which the procession of Saint Benedict and Saint Sebastian passes.

This itinerary presents itself as a space/time of the sacred, as Rosendhal (2008 a,b) points out, as well as composing a conniving landscape, a mark and a matrix, as Berque (2004) points out.

I perceive a giving, receiving and giving back in the same sense as the network configuration of the social movement that makes the Ticumbi celebrations effective, through faith and solidarity, but also a political action (understanding the meaning of social action from WEBER, 2004).

I also observed this network of affections in the sub-institution of specialists in healing, namely the midwife, the benzedeira and the rezador. What is exchanged during a blessing? Magic words are exchanged, as Caillé (2002, p.99) shows us, in a ritual evocation of the sacred in order to break a copperiro, an eye (of good and evil), cure a fallen spine (emic diseases - of the place, local), since evil is cut, broken and cured. In this case, what is given and what is received are in the category of the sacred and follow the ritual of the Gift, of the bestowal, since "it is precisely the bestowal, that is, the set of services provided, not with the aim of acquiring a more useful *good* than the good given, but to seal a *bond*; moreover, this operation occurs with the denial - at least temporarily - of the rule of equivalence" (CAILLE, ibid. p. 105).

These healing specialists are mediators between the sacred, the saints, the gods and men. They receive a *divine gift to heal, to make a birth happen, to solve a difficult birth, to bless, to remove a spell, etc.* We'll look at this in more detail in the next chapter.

6. Healing know-how in the village of Itaûnas: an expression of spatiality.

Fàbula of Higino

Once, while crossing a river, Careful saw a lump of clay soil: he thought about it, took a piece and began to shape it. While he was reflecting on what he had created, Jùpiter intervened. Careful asked him to give spirit to the clay form, which he gladly did. As Careful then wanted to give his name to what he had shaped, Jùpiter forbade it and demanded that he be given his name. While Careful and Jupiter were arguing over the name, the Earth (tellus) also appeared, wanting to give its name, since it had provided a piece of its body. The disputants took Saturn as their arbiter. Saturn pronounced the following decision, apparently equitable: "You, Jupiter, for having given the spirit, must receive the spirit at death and you, Earth, for having given the body, must receive the Body." Since it was Care that first formed him, he must belong to Care for as long as he lives (Heidegger, 1995).

When relating the healing practices and knowledge of the specialists in the village of Itaûnas, ES, with the spatiality engendered by them, it is worth noting that human societies considered to be *slow* or with a *traditional* way of life have always sought alternatives based on endogenous action logics and not always based on the proposals of hegemonic institutional health logic (OLIVEIRA, 2000). In this process, these societies assume a relationship with their surroundings, looking for elements in the environment that integrate both mythical and magical aspects of healing, as well as taste, color, appearance and smell, allied to a given lunar season in their calendar. These elements are promoted to curative materials through ritual processes where the word of prayer, intonation and chanting exert a strong curative impression, symbolic efficacy, according to LÉVI-STRAUSS (1970, 1975). This type of society/nature relationship builds up a certain *body of* empirical/magical knowledge and, therefore, its own spatiality marked by existence, in a *broad* sense, and by interpersonal relationships and their environment, in a strict sense. In this know-how, with an organized *corpus* of knowledge - concrete thought according to Lévi-Strauss (1997) - these societies weave a web of relationships of meanings and signifiers that operate within a system of classification legitimized by their *worldview*, where everything has a meaning, finds an answer. These symbolic representations of a *modus vivendi* are perceived as expressions of a *spatiality* - an ethos, a culture, thinking of the notion of Heideggerian *dasein*, and in the sense given to it by Holzer (1998, p. 40) - a *dasein* permeated with spatiality, geography, semioticians who geo-graphize a world. This *body of* knowledge of the art of healing is a local social strategy for dealing with the health/disease process in order to restore well-being, balance and healing. The body is not seen as a system isolated from the world, but in *relation to* the *whole cosmos*, and in relation to others in the world - otherness[85] - intersubjectivity[86] -. As Ayres (2001, p.64, 66-67) put it: *I am what I see of myself in your face. I am because you are.* (Proverb from the Zulu tradition - South Africa).

Within this system, everything makes sense, the macrocosm and microcosm are connected and the world obeys a gnosiological order that comes from a religious tradition and faith. These strategies are not isolated from

85 It refers to the experience of the relationship with the other, otherness, in which I am constituting myself, ipseity (Ayres, 2001).

86 As Ayres (2001, p.64, 66-67) put it: *I am what I see of myself in your face. I am because you are.* (Proverb from the Zulu tradition - South Africa)

biomedical knowledge, but are articulated with it, as we will see below.

It can be argued that in the worldview of these slow societies, nature is seen as sacralized and active, natura *naturans* as opposed to *natura naturata,* passive and a mere resource (SODRÉ, 1988, p. 152). In the sense of this living nature, Godelier (2001p. 160-161) says that:

"(...) The entire universe is no longer composed of anything but people (human and non-human) and relationships between people. The cosmos becomes the anthropomorphic extension of men and their societies. The individual is linked to the entire universe, which surpasses him and contains him and also surpasses his society."

In this anthropomorphic cosmos of Godelier (2001) or of bioanthropomorphic beings in the sense of Diegues (1996: 54) and Morin (1999: 195), there can be communication and communion between human beings in a state of community (MOSCOVICI, 1988: 56) and the world (material and immaterial). Therefore, for these communities that still maintain traditional ways of life, the territory has a strong symbolic value, as it is the basis for the formation of group/individual, political and religious identity, important in the recognition of oneself by others.

6.1 The chegança in Vila de Itaùnas and ethnographic work: some considerations.

It's not an easy task to arrive in a community of belonging and feel at ease. Especially because overnight I changed from the category of 'tourist' to 'writer'. I gained some unexpected visibility, and from being an *observer* I became the *observed.* Doubt hung in the air and everyone wanted to know who I was? What was I doing? For what? For whom? Why? All I heard were questions. I'll try to answer these questions here.

In the summer of 2006, I decided to carry out my fieldwork in the village of Itaùnas, Conceiçao da Barra, given the ease of access, the small population, and the fact that I already knew the village, I thought that fieldwork and participant observation would be more productive. There was also the fact that I had been a regular in the village since the late 1980s and spent Holy Weeks, Carnivals, my best holidays and moments of leisure and relaxation there. This 'choice' also came about because the Vila's territory was redefined as a Conservation Unit for restricted use, and (re) territorialized by the State through the installation of the Itaùnas State Park. Knowing the history of the village, at least in part, I could see through the trained eyes of Geography, even if I couldn't see any discord, that there was a certain conflict in the air. Redefining territory in an already constituted communal territory generates conflict. I tried to create a piece of work that would show that there was a spatiality there, a different *modus vivendi*, a cultural territory in its living, dynamic sense, and not just a political and institutional demarcation (in the hegemonic sense). There had already been a previous, very successful experience in a work on the Saco do Mamangua community, which made it easier to see certain invisibilities - a work of phenomenological hermeneutics.

The first forays into the field lasted a few days, were quick entries, where it became clear that there was a mutual estrangement (community x writer) and an inadequacy in relation to the first contacts, which was to be expected. The second visit took place in the summer of 2007, and I stayed for fifteen days, returning for another fifteen days. In the winter I returned more quickly to confirm some data. In the summer of 2008, I completed my work with participant observation at the complex ritual[87] , and its developments in commemoration of Sao Benedito and Sao Sebastiao, on January 18th, 19th and 20th.

This latest fieldwork changed every conception that had existed with regard to the social relationships and configurations that make up spatiality in Vila de Itaùnas. As a result, all the interviews were revised, the notes redone, completed, and an interpretative map was constructed based on a mental map, with the help of the recorded short films, the interviews, in short, all the iconographic documentary material was fundamental to the construction of a new theory on spatiality in the village. Given the path and the wayfarer, this ethnographic work is sufficient for the moment, since it would require a second phase of research to expand on it.

In this total change, I decided to present the spatialities in the Vila as Seemann's (2008) carto-facts, with the visible and invisible landscapes as texts. The concept of the conniving landscape (BONNEMAISON, 2002, CORRÊA, 2004, p.86) helped us a lot, giving us an understanding of its mediating quality, in other words, a landscape that has visibility and, at the same time, brings with it invisibilities connected to the hidden world of affectivity, mental attitudes and cultural representations, which are woven like a mesh through everyday relationships. These relationships were the focus of this work, since it was in this intricate mesh permanently produced on a daily basis, through the participants of the religious and political institution of Ticumbi, who are at the same time and dialectically the inhabitants of the village, that I found the basis for the ethnographic interpretation.

To get to Vila de Itaùnas, you have to overcome the challenge and take a dirt road on the left-hand side of the highway from Conceiçao da Barra to the BR 101. This road leads through a small patch of Atlantic forest and a huge eucalyptus grove to the Vila de Itaùnas district. At various times it is possible to get lost if you don't follow the right track, so you need to be careful along the way. On rainy days, the road is muddy and dangerous. As soon as you arrive in Vila, you can see a new neighborhood on the left: the expansion of Vila. A new subdivision that was created following a claim by local actors, who needed new land to build on, since they had sold their houses to the new residents - outsiders, tourists and their second homes.

When you get to the village, you see a small village, with people wearing flip-flops, simple clothes and walking at a calm pace, and it's as if you're in another time/space. There are no buildings, only small houses, at most townhouses. The houses of the local actors are small, colorful and generally have two bedrooms, a living room, a kitchen and a bathroom. All the houses have backyards where they grow plants for medicinal purposes, some vegetables, fruit trees, and some keep chickens. Some houses have small buildings in the backyard, *suites* for renting out to tourists and guaranteeing an extra income. There are many guesthouses, two small supermarkets, two bakeries, a pharmacy, a police station, a secondary school, a health center, a nursery school, and many restaurants and small and medium-sized bars scattered throughout the central part of the village. There is also a fishermen's association next to the police station and a cultural center where women make handicrafts.

The road is unpaved and sandy, and when it rains, large pockets of water, sand and mud form, making walking

87 Ja referred to by Alves (1980), when he uses this term to emphasize a set of rituals or ritual sequences, unfolding in events that combine the same elements expressed spatially in rites of faith.

more difficult. The most used vehicle in the village is the bicycle[88] and its use is encouraged by the park manager. For this reason, the village is quiet and calm outside of festive and tourist periods. Its inhabitants like forró and sertaneja music and are used to listening to music on the radio or electronic devices. The children are cheerful and play in the streets and in the square. The street corners and squares are the favorite places for children and young people, and in the late afternoon and early evening you can see them on the benches in the square, or gathered in groups chatting on the street corners. Hά a volleyball net in the square, where young people have fun in the summer. Hά a well-kept soccer field where local soccer matches take place.

On sunny days, boys and girls often bathe in the Itaûnas River and it's one of the best sights to behold: the river with its dark water, the children frolicking and having fun in the water, the dunes in the background, and a blue sky framing it all.

The village has a central square, where you'll find the parish church of São Sebastiao, a flagpole with the saint's flag, and a huge fallen pequi vine trunk, three important geosymbols that point to the history of the place. This isn't the only church, because down the street, next to Dona Tereza's restaurant, you'll find the unofficial Igreja de Sao Benedito. There is also an Assembly of God, which is open on Sundays and Wednesdays, a Baptist Church (services on Sundays and Wednesdays), a Maranatha Church and a Maranatha Renewed Presbyterian Church. Although religiosity in Vila is predominantly Catholic, the number of Pentecostal believers has been growing, which causes some discomfort for the Catholics. The Pentecostals don't take part in Ticumbi, but neither do they engage in rivalry. In the interviews, they showed that they prefer to keep a low profile in terms of their religiosity, *each with their own religion, their own custom*, according to one interlocutor. At the end of the village, next to the River Itaûnas, is the headquarters of the PEI, and following the bridge, next to the dunes is Aldeia Gregório, an artisanal fishing village within the PEI. This is a difficult negotiation, as the PEI is a restricted-use CU.

Summers in the town of Itaúnas are hot, with a scorching sun, lots of tourists and lots of traffic every day. In relation to the sun, you need to be very careful about the times you arrive and leave the beach, as crossing the dunes between 11am and 3pm is not healthy. I've had some disastrous experiences in this regard, as I've had to interview fishermen when this is the time they're returning from their fishing trip, collecting their fish and nets. In this sense, I looked for milder times to visit the interviewees in the village, giving preference to the afternoon, or tardinha, as they say. At these times in the village, people are more relieved of their domestic chores; the men are at the doors of the houses or shops chatting, the women have already bathed their children and are going to church to pray, to a friend's house, or exchanging information with the neighbors at the doors of the houses.

The village wakes up early, people wake up and start their chores straight away. In this sense, I see people building paths, tracing their daily itineraries. The artisanal fishermen ride by on bicycles[89] on the *fisherman's*

88 The poet and actress Elisa Lucinda lent me her bicycle to observe the Vila. I would like to thank her for her kindness, company, encouragement and poetry.
89 These bikes have a box (colored, sometimes red, green or black) attached to the back or front, and they pass very quickly. They usually ride in groups of 4, 6 or more.

trail[90] , this scene is one of the most beautiful, as it is very reminiscent of a procession, a procession to work, the daily toil. On the way back, in the afternoon, they follow the same route.

At night, outside of the tourist season, there isn't much to do, and in general people don't leave the house. Except on weekends when there's a forró activity, a birthday, or rehearsals of the Ticumbis. These rehearsals either take place in the village itself, or the congueiros are invited to a nearby location. The rehearsals are very lively, like a party. The owners of the house who invite the Ticumbi group host a reception with drinks and food for everyone. Often there's a little family dance that goes on well into the night. Everyone takes part in these activities, from children to adults and the elderly. It's a get-together for everyone. This is a preparation for the big celebration in January, the month of new beginnings in the village. The year doesn't start until after the festivities of São Sebastiao and São Benedito," said one interlocutor. This reminds me of the understanding of this festival as a great liminal ritual, a state of passage from one year to the next (TURNER, 1974).

Festivities are a moratorium on everyday life and are always necessary and repeated (MARQUARD, 1993), because festivals mark time and space. Festivities invent and reinvent forms of identity, at the most basic socio-geographical level (DI MÉO, 2001). It is necessary to have a party, to celebrate life, to thank the saints, to break with everyday life in order to start again, to innovate, to be able to reinvent life.

The days of calm come to an end during the festivities (and the sightseeing), and the village is transformed into a cinematographic scene, as if a show were being presented every hour. It takes a lot of disposition and physical stamina to keep up with everything. But for faith and through faith, the devotees make a point of taking part and not missing a single event. Although the festival is a very tiring ritual, the older representatives make a point of taking part in every event and every day. They only go home to bathe and sleep, the rest of the days are spent in the festivities that are organized, such as: the procession of Sao Benedito, with the arrival of the saint (the image) by the river Itaûnas, the presentation of the various Ticumbi groups in the two tents (of Sao Benedito and Sao Sebastiao, see the interpretive map in chap. 4, p. 60), the presentation of the Alardo, the presentation of the Reis de Boi groups, St. Sebastian's mass and St. Sebastian's procession through the streets of the village and back to the main church.

It is a true communion of solidarity, a complex ritual that dissolves patterns of social hierarchy, where these actors remain for three days in a state of *communitas*, a term coined by Turner (1974, p.119):

> "I prefer the word communitas to community, in order to distinguish this type of social relationship from an "area of common life". The distinction between structure and *communitas* is not just the familiar distinction between "mundane" and "sacred", or that between politics and religion, for example. Certain fixed positions in tribal societies have many sacred attributes; in fact every social position has some sacred characteristics. However, this "sacred" component is acquired by the beneficiaries of the positions during "rites of passage", thanks to which they change positions. Something of the sacredness of transient humility and modellessness takes the lead and tempers the pride of the individual entrusted with a position or higher office. (...) It is rather a question of recognizing an essential and generic human bond, without which there could be *no* society. Liminality implies that the high could not be high without the low existing, and those who are high must experience what it means to be

90 A path through the restinga, already in the park area, leads straight to the itaûnas, the rocks in the sea. *It's a marked* path that you enter with your canoe.

low."

During these celebrations, we find social positions changed, as we saw in another chapter. Artisanal fishermen and simple farmers become Bamba kings and Congo kings. The simple girl from the outskirts of Vitoria becomes the queen of the congo, the merchant is the sailor, the vassal, the warrior of the Alardo. In short, this ritual presents and re-presents social positions. "The experience of each individual's life makes him alternately exposed to structure and *communitas*, to states and transitions" according to Turner (op.cit., p.120). The festival ritual transforms the common places of Itaùnas into sacred places, marked by strong rituals of great cultural identity expression. This is where the territory appears as the *embodiment of culture* (CORREA, 2004); in this sense "the territory favors the exercise of faith and the religious identity of the devotee" (ROSENDAHL, 2008, p.57).

6.2 Popular Catholicism in the town of Itaùnas and devotion to the saints.

The healing know-how in the rural community of Vila de Itaùnas is organized within the Ticumbi Institution, a popular Catholic religious association that worships two saints: Saint Benedict and Saint Sebastian, who are honored on January 19th and 20th as a local cultural tradition. St. Sebastian is the patron saint of the town of Itaúnas and St. Benedict is one of the most popular saints in Brazil. The latter is worshipped mainly in rural Brazil and by the Afro-Brazilian culture, and in the village the social actors, especially the specialists in the art of healing - the benzedeiras and the rezador - owe him special devotion. In the words of the 81-year-old benzedeira: "I pray a lot to St. Benedict, and there isn't a day that I don't pray to St. Benedict!"

As previously mentioned, St. Benedict, like St. Gonçalo, is a party saint, his homages are always with a lot of partying, dancing and joy - he is a playful saint, in the sense of Zaluar (1983). Unlike other saints, there is no set date for celebrating Saint Benedict:

St. Benedict is our patron saint. St. Benedict is celebrated all year round .[91]

The saints are the mediators between God and man on earth and they have a special importance within popular Catholicism, as we shall see.

Religiosity is an act of recreating the world, and it is always a collective undertaking carried out with the social and historical resources of a given culture, in other words, we weave the uniqueness of our *modus vivendi* with the threads we have inherited from the past. By recreating the past, we invent the new, always a reinterpretation, framed by the socio-cultural aspects of our *place*. In this sense, our Catholicism has strong Iberian roots, since along with the images of the saints, brought by the Portuguese colonizers, also came the beliefs and myths that gave rise to Brazilian popular Catholicism. However, although we have a hybrid culture between three ethnic groups: Indians, blacks and Europeans, the indigenous influence on the elaboration of our religious cultural tradition is less evident than that of Europeans and Africans, according to the scholar religiao Steil (2001, p. 14).

In order to understand this religious process, it is necessary to note that the Brotherhoods emerged in Brazil

91 Fragment of the lyrics from Jonathan Silva's song, CD - Benedito (2007)

between the 17th and 19th centuries. According to Steil (op.cit., p.18-20), the Brotherhoods are groups of lay devotees who have organized themselves as private, non-ecclesiastical associations whose aim is to maintain a cult or devotion. Even though they depend on the clergy to perform some rituals, these associations maintain their autonomy from the Catholic institution in legal and economic terms. Until the arrival of religious congregations in Brazil, all urban sanctuaries were run by the Brotherhoods, which owned the temples and were responsible for maintaining religious worship. The expropriation of the Brotherhoods, followed by their replacement by clerical religious congregations, took place through struggles and conquests, at the cost of conflicts marked by strong symbolic and police violence. However, the Brotherhoods resisted and managed to maintain control of their sanctuaries, while the monks, blesseds, pray-ers and benzedeiras lost their control over the rural sanctuaries and local chapels, and their activities were restricted to the private spaces of homes, on the bangs of the Catholic institution, under the auspices and legitimacy of non-clerical religious associations. Following this line of thought, we can infer that the institution of Ticumbi, in the village of Itaùnas, is similar in its organization to a Brotherhood. The Brotherhoods were the most consolidated and the most resistant to the process of removal of religious power by the clergy. In the second half of the 19th century, lay leaders were replaced by priests from religious congregations as part of a project to modernize Catholicism with a clerical model.

The mission of these modern agents who came from Europe would be to purify traditional or popular Catholicism by combating superstition - an undertaking of the new civilizing mission (STEIL, op.cit.). It was this intervention that would demarcate the division between traditional popular Catholicism and the enlightened clerical Catholicism of the religious system as a whole. For Steil (op.cit), it will be in the shrines, such as Aparecida do Norte, among others, that we will see this tension between the segments, since the strong presence of religious congregations in the direction of the shrines had already consolidated a sacramental religiosity that was absorbed, incorporated and reproduced by Catholics in Brazil. There was a great deal of discrimination against the Brotherhoods, which came to be seen as a stage in the religious past, but which should disappear with the passage of time. But they didn't disappear, as the two forms began to articulate with each other. In such a way that the elements of one were incorporated and re-signified by the other, in a circularity where one form feeds the other (STEIL, op.cti., p. 17).

In this traditional system of Catholic religiosity and in rural areas, at the cost of much resistance, the local beatos, monks, pray-ers and benzedeiras became responsible for the day-to-day maintenance of the beliefs and rituals of faith (STEIL, op. cit., p. 19). The priest was generally a distant figure, who visited the shrines on the occasion of the patron saint celebrations once a year, the procession and every two years for weddings and baptisms (STEIL, op.cit).

With this same socio-geo-historical and religious configuration, we will find a rural society in a state of *communitas*[92] , such as families of fishermen and small traders in the village of Itaûnas, Conceiçao da Barra, ES in their religious rites.

92 A term used by Victor Turner (1974, p.118) to refer to a society in a state of community - a communion of equal individuals submitting together to the general authority of ritual elders.

In a conversation with a traditional family in the village, a couple aged 81 (woman, E1) and 82 (man, E2), husband and wife respectively, and their 44-year-old son (man, E3), I realized from their statements that the priest was a figure far removed from the local reality:

E1- But the priest who comes to say mass here is from Barra, you know... Conceiçao da Barra! He comes in the morning to say Mass for Saint Sebastian. In the afternoon he's not even there! He leaves!

E3 - go away... it doesn't solve anything anymore.

E2 - Lazy... I say he's a lazy priest!!! He barely... barely comes, says mass, disappears, my lord and leaves... doesn't even attend the party!!![says indignantly]

(interview with the family where E1, wife, E2, son, and E3, husband, January 2007,)

According to this couple, in the old days (a time/space in the memory of Vila antiga), in the absence of a priest, anyone with a *suit* could celebrate a wedding or a baptism. As long as they had a suit, and there was faith, according to an 82-year-old man, because in the 'old days' the priest didn't have a habit of visiting the village and saying mass regularly. In this case, the community would sort itself out and the priest could be replaced by someone wearing a *suit* (like the uniform[93] used to officiate at the event), which gave him authority, and faith in God and the saints, which gave him social legitimacy.

In popular Catholicism, saints are identified with images (statues), which are iconographic representations of these entities, which are both people and divine spirits that exist locally. Devotion to images is central to popular Catholicism, and they are in fact the place where the invisible becomes accessible and palpable (STEIL, 2001).

In the same way that human bodies are the repositories of invisible souls, images are the bodies of saints. Through images, communication is established between the living and the dead. Based on the dogma of the communion of saints, this model of Catholicism creates a cosmology in which the boundaries between life and death are continually crossed without necessarily the mediation of specialized agents. The relationship between the saints and the faithful is personal and based on the principle of protection and loyalty. Each believer has their patron saint, or their heavenly godfather, who in return asks for their loyalty. Many scholars of Brazilian culture have shown how this relational model not only served as the basis for legitimizing relations of domination in Brazil's manorial society, but also remains today as a long-standing cultural element that underlies the relations of patronage and clientelism that are still so present in Brazilian politics today. This relationship between the saint and the faithful can take two forms: alliance and contract. The covenant relationship begins at birth, when the person is consecrated to a heavenly godfather, creating a lifelong commitment between the two. (...) The contractual relationship is associated with promises and pilgrimages to shrines. The saints in the cosmology of traditional Catholicism generally have their own specialties. (...) In moments of crisis, the faithful make their requests to the saints, promising them a sacrifice in return for the favor received. In this way, a system of exchanges of symbolic goods is established between the living and the dead, generally narrated as miracles, involving the faithful and the saints in the same linguistic community of meanings (STEIL, 2001, p. 21-22).

These images can belong to the Church and religious brotherhoods, such as St. Benedict of Ticumbi, and Our Lady of the Brotherhood of the Good Death (Corrêa, 2004), in which case they would 'live' in local chapels or

93 Da Matta (1997) differentiates between uniform and costume. Uniforms, as formal attire, operate through individualization or in an analytical way, rigidly and clearly segregating one role from the others. The uniform refers to a given position within the social configuration, since it is a symbol of power in the social order. In this case, the *suit* invests the social subject with power, so that they can carry out their duties.

churches, but they can also belong to individuals or families, and can therefore 'live' on altars specially built inside their homes (ZALUAR, op.cit, p. 59). Catholic saints are the saints of the social order: they are the saints of legitimate illness and legitimate healing. Those who 'guard' the image of the saint have power and legitimacy, and believe they have the saint's protection, but also very serious obligations, both in relation to the saint and to his devotees.

It is the saints who represent the various networks, categories or groups of people organized in everyday life. They are the ones who bless important passages in the life cycle of individuals and in the cycle of agricultural reproduction. Social control in Catholicism, according to Zaluar (op.cit.p. 114), operates through a code, which includes both the control carried out in the networks of solidarity, the networks of affection, activated by and through the various religious practices, as we have already seen in the case of the feasts of St. Sebastian and St. Benedict, and the power of the priest and other church officials, the result of their knowledge and class positions.

In this way, the image of a saint is not just a representation that evokes someone who was in the world of the living, but it is "a sacrament" - hierophany[94] - something that is made present in the visible world, in an effective and real way. The relationship between the image and the saint makes them one and the same, and for this reason, places and images have a special, particular meaning in popular Catholicism, a singularity that goes beyond any attempt at rationalization or generalization (STEIL, op.cit.). Eliade (1996, p. 7) tells us that: "the symbol, the myth, the image belong to the substance of spiritual life, that we can camouflage them, mutilate them, degrade them, but that we can never extirpate them." In the Brazilian countryside, as well as in the village of Itaúnas, I found cases[95] of saints being stolen from churches, and saints being transferred into the hands of 'families' who would be the guardians of the saint .[96]

There isn't a single popular Catholicism, but several, since the ecclesiastical Catholicism or that of the official Church is not the same as rural Catholicism, or that of the urban proletariat and other social strata in Brazil (ZALUAR, op.cit., STEIL, 2001). These Catholicisms differ mainly in their practices and devotions, beliefs and *modus operandi*. In this sense, there is a coexistence full of tensions between popular Catholicism and ecclesiastical and hierarchical Catholicism, or as they say in popular language: the Church of the people and the Church of the priest. In Vila de Itaûnas, this division is very well represented by the Church of São Sebastiao, which is official, and the Church of São Benedito, saint of the people, of the blacks, of the workers - and not official, as indicated by a social actor in the village:

"That little church there... have you been there?

Nearby... in front of the forró bar?

---

94 Conf. Eliade (2001), as well as Rosendahl (2001, 2008 a, b) on the territory of the sacred.

95 That's what they call the cases of the disappearance of images. On the day of the São Benedito festival - after the Ticumbi performance - I sat next to the benzedeira and her 'white son', a resident of Vitoria, who told me some of these stories about the disappearance of a much-loved image of São Benedito from the town's Mother Church. "I think Saint Benedict is going back to Itaûnas!", said the young man. He was talking about the image as a living being, not a dead one, or just an image! As referenced by Steil (op.cit.) and Zaluar (op.cit.).

96 See Gilberto Freire (1933) and Zaluar (op.cit.) on the social order and the power of male domination in Catholicism.

There's a little church like this.... in front of the forró bar... There's a little church.

Only there's a cross like this.... that's the little church."

(interviewee, 44 years old, 2007, indicating the location of the unofficial church of Sao Benedito, in the village of Itaúnas)

Even the term 'popular' is in opposition to the erudite culture of the priests, and is a way of marking the difference and a certain disqualification. The so-called popular medicine and official medicine are in the same position. However, in this book I will call it the knowledge of the art of healing of the specialists in healing, organized within the religious association called Ticumbi.

In this sense, I'm opening a parenthesis to include a discussion of the knowledge of the art of healing, discussed in Xavier's work (2004). In national and international literature, in common sense and in public bodies and institutions, this knowledge is known as *traditional medicine*, *popular medicine*, *folk medicine*, *popular health culture* and *ethnomedicine* (Büchillet, 1991, p.24). However, the writer Delma P. Neves (1984, p.7) argues that the above terms operate with ethnocentric views because they take erudite or scientific knowledge as a reference. For this author, in explaining popular medicine as 'residual' or 'alternative', or even as 'parallel to scientific medicine', many authors are unable to get away from the comparison or confrontation between these two social practices. According to Neves (ibid.), this practice/knowledge is perceived in the West as residual or alternative, and is therefore expected to be eliminated as soon as economic and social 'development' takes place; while popular medicine is defined as the magical-religious knowledge of rural communities or the subaltern layers of urban-industrial societies (BOLTANSKI, 1989; MOTTA-MAUÉS, 1993). However, this practice is not understood as "medicine" by its practitioners:

"So much so that the classification of folk medicine is an external qualification attributed to this practice, usually by researchers, to differentiate it from scientific medicine. The users and agents who consider themselves to be custodians of this knowledge accumulated over generations and who specialize in prescribing such medicines do not consider it to constitute medicine." (NEVES, ibid., p.7)

In general, these social actors don't equate the two medicalization practices as alternatives that can be generically substituted for each other. And if they do, they recognize the limits between the two types of knowledge and practices and their respective *modes of operation*. Even though they use traditional practices, they always demand greater attention from the state in the provision of medical services and in controlling the price of industrialized medicines.

For Neves (ibid.), the way in which these social actors, users of *home remedies or bush remedies, construct* their discourses has to be qualified and contextualized. This is at the risk of externally attributing positive (as well as negative) and very broad (or distorted) values to the knowledge about the body and diseases, on which this practice is based, and whose limits they themselves recognize. For the aforementioned author,

"The term "*folk medicine*" can only be accepted if we agree with an evolutionist (cf. Foster, for example) or diffusionist (cf. Boltanski) interpretation to explain the existence of therapeutic practices based on knowledge of the healing capacity of certain plants, which men have built up in their experiences of dealing with the problems posed by illness. If we want to break with these perspectives, we must also abandon the term medicine and, therefore, popular medicine and the comparison of these therapeutic acts with (scientific) medicine as a presupposition for understanding their specificities." (NEVES, ibid, p.11-12)

In agreement with Boltanski (1989), Neves (ibid., p. 10) points out that biomedicine's explanations are based on a single way of conceiving relations of domination, i.e. of one class over another, ignoring and/or not taking into account the forms of domination and the relations of power that occur between segments and agents of the same *social class*. Social actors such as shamas, healers, benzedeiras, midwives, pastors, among others, occupy a *'social space'* within the order/class and therefore exercise power within it. This makes all the difference when we talk about the traditional population, as well as the so-called *'classes populates'*[97] . The specialists in the art of healing, benzedeiras and rezadores, as mediators between human beings and the saints, within the social order fulfill the role of healers of the legitimized illness, establishing the legitimized cure.

6.3 Benzedeiras and rezadores: some theoretical aspects and the reference in the village of Itaùnas.

These specialists in the art of healing are considered within popular Catholicism to be human mediators, according to Zaluar (1983, p. 103) and Steil (2001, p. 24). They occupy a social *place* in the community through religiosity and local politics as mediators between the saints and human beings - inserted in a fine mesh, and exercising legitimate power through their knowledge. Through the work of faith, they meet the needs of the faithful through the rites they perform in their homes, and it is through the exercise of living faith that they reproduce and give meaning to the world of the devotees, restoring a balance in the social and symbolic world. In the local ideology, it is the saints who guarantee healing, despite the symbolic effectiveness of the prayers and blessings used in healing rituals. Symbolic efficacy is a concept developed by Lévi-Strauss (1975) in order to explain the phenomenon of shamanistic healing. It results from the possibility of manipulating the organs of the human body by means of symbolic rituals, i.e. equivalent signifiers of the thing signified (the organs). Through prayer, a language is created that establishes this relationship between the symbol and the thing symbolized, thus allowing the expression, in a realistic order different from the thing symbolized, of states and conflicts that were previously unresolvable and leading to their resolution on a symbolic level (ZALUAR, op.cit., p. 103, note 8).

The power to heal is associated with faith in the saints, as has already been said. Each saint was associated with different parts of the body, because each one was in charge of a part of the body and their power over it was absolute. According to Zaluar (op.cit., p.104), the charisma of the healers lies in the symbolic power of the saints, whose help will be indispensable, both in healing and in legitimizing the healer's role in the eyes of the other members of the community. The author adds that these specialist healers were attributed charisma[98] , or symbolic power from the community of people they represented, due to their exemplary behavior, evaluated within a specific local ethos and morality. The positive values that underpin this practice, and the actions of these specialists, are supported by the attitudes of humility, suffering and poverty assumed by the individuals, at least for the purposes of externalization.

In a brief discussion on the subject, at the limit, I can say that unlike official medicine, this is an art of curing illnesses legitimized by the social order, in a religious process of popular Catholicism. In this way, I propose that this experience be understood as a religious experience, just like the experience of art, beyond the aesthetic

97 Cf. Boltanski, 1989.
98 Weberian charisma.

experience, as configured by Gadamer's (1985) philosophical hermeneutics[99] . For Gadamer (op.cit.) "the being that can be understood is language". And Language for this author will be everything that can be expressed, can be said, and not said, but understood. The texts of this language are beyond linguistics and in this sense hermeneutics helps us, since it broadens this phenomenological gaze, illuminating something that has not yet been illuminated.

Seeing and hearing beyond a truth, through experience. This experience of Gadamer's art (op.cit., p. 38), the essence of this experience can be understood through play, play as in Huizinga. Let me clarify that play in German - *spiel* - can be the performance of a musical piece, the acting and interpreting of theater or cinema. The spectator, as a participant in this process (game) of understanding the work of art, can or should understand it in this openness of possibilities in which the work (language) as a thing offers us itself, or presents itself. This is the experience of art. Transcendence takes place in this experience.

In this sense, I can infer that the art of healing, being art, in its act, gives rise to an opening of understanding, where the spectator or devotee - the other, enters into this understanding mode and understands - including themselves (as being-ai *of dasein*) through the ritualized dramatization of the healing specialist - the healer, the pray-er, the shama. In fact, they both understand and transcend a truth, a rationality, while at the same time allowing this unity, this rationality, to continue to act. This unbalanced social subject finds meaning in something that was outside the cosmic order - the illness, the disorder, the malaise. He actually lives an 'experience of art' through a religious ritual such as a blessing. Why is this possible? How is it possible?

Heidegger (2002, p.165) says that the being inhabits the world, and inhabits it poetically and playfully, and for this reason distinguishes itself from other beings. The being in the world poetically spatializes and worldizes, since it gives meaning to everything it touches. These meanings belong to their cosmology, the cosmic order that is expressed through their ethos, their *modus vivendi*. What do the officials of the art of healing do in the rituals, through their poetically linked prayers - a *modus operandi*, other than to lend, through and by art, a meaning that organizes the disorder - illness, so that through this game, they can operate transcendence and re-establish order with healing?

Without wishing to delve deeper into the issue, given the limits of this work, but taking advantage of the opportunity, I use Ayres' reference (2001, p.66) in relation to Kant's allegory, used in another interpretation by the southern philosopher Stein (1976, apud. Ayres, ibid.) on the flight of the dove:

The light dove, while in its free flight, cuts through the air whose resistance it feels, could imagine that it would be even more successful in the vacuum.

I have already discussed this subject in another book, but I insist on the allegory. For Ayres (op.cit., p.4) the dreamy flight of the dove can be understood as a strong metaphor for the identifying act, in the attribution of predicates to the different moments of experience that make us constitute our worlds and ourselves, in a reference to Heidegger (1995). According to Ayres (op.cit.), "this refers to the process of *constructing identities*, which shows us an inexorable dialectic of denying by constructing/constructing by denying, so

99 Hermeneutics as the art and technique of translating, leading to understanding.

difficult to put into words and so clearly expressed in the Kantian metaphor". In this author's dialectical interpretation, the place of the subject as a thinking being, who by attributing predicates to the world, identifies himself. We are interested in the experience of "resistance" that makes these other subjects emerge as real presences, because it is in the friction with the other (otherness) that I constitute myself (ipseity). The free flight of the dove can be understood as human history itself, the resistance of the air will be the inexorable presence of the other, as referred to in Ayres' article (op.cit.). It is through the act of spatializing the world, in this friction with others and with the things of the world, in this relationship, that the being, the social actor in Vila de Itaûnas, semiographs the world, at the same time as he is in the process of identifying it. And where is language? Ayres (op.cit., p.68) says that :

"(...) there seems little doubt that one of the most powerful means of this mutual engendering of subjects and worlds **is language. It is from language that comes the fundamental 'resistance' that puts us in the presence of the other**. Heidegger (1995) had already stated that language is the dwelling place of being. Gadamer (1996) added that being that can be understood is language. Now, just as air does not come into existence as a world until it is experienced by the dove as resistance to flight, so language only exists as an act (Habermas, 1988). And what is this act, which creates subjects and their worlds in and through language, which makes us experience what we are in the encounter with what we are not, if not dialog? We can then say that subjects are dialogues" [emphasis added].

I would venture to say that it will be through language (beyond linguistics) between beings, a poetic expression (because poetically the being inhabits the world, said Heidegger, 2002), which operates through *play* (Gadamer, 1985) without any pretense of truth, a transcendence, in which, as in the *experience of art,* a cure can occur. Of course, I'm not going to question the therapeutic value of plants, but in the case of the blessing, it's mainly through the transcendence of faith, of the devotee's religiosity, that there were magical and poetic words, that a cure occurs, a well-being, relief for the afflictions of these beings in a *state of community* (MOSCOVICI, 1990).

The main emphasis in the relationship between these *men and women of God* and their believers is generosity - since they don't charge for consultations, prayers or prescriptions for bush medicines, baths, etc. Their knowledge is passed on, used as a *gift* - like Mauss' *gift of reciprocity* (1974). They have received the gift from God and make their counter-gift offer to the saint, exercising their function in the art of healing the devotees, as a religious service, a service of faith, through faith, and for faith. In this sense, they resignify a being in the world, a *dasein*, expressing their *modus vivendi*, their internal relationship of values externalized in a social space.

This *modus vivendi* assumed by these figures of popular Catholicism, with their extraordinary powers, is understood here, as in Zaluar (op.cit), as symbolic power. Symbolic power for Bourdieu (1998) is this invisible power which can only be exercised with the complicity of those who don't want to know that they are subject to it or even that they exercise it.

Symbolic power is a power to construct reality that tends to establish a gnoseological order: the immediate meaning of the world (and, in particular, of the social world) presupposes what Dürkheim calls *logical conformism*, that is, a homogeneous conception of time, space, number, cause, which makes it possible for intelligences to agree. (BOURDIEU, 1998, p. 9)

There is an important distinction between the benzedeira and the rezador in the religious world of popular

Catholicism, which is also reproduced in the village of Itaûnas. This issue is related to gender and the distribution of religious services within popular Catholicism. Zaluar (1983, p. 112) informs us that social life is marked by certain passages, and in order to enter and leave them, it is necessary to go through a ritual. These rites of passage constitute the undisputed religious competence of the figure of the 'priest'. His knowledge consists of prayers, the use of holy oils and other Catholic symbols that legitimize these fundamental passages in the life cycle of any social being. In reference to this knowledge of the priest, there are the functionaries - the pray-ers and benzedeiras who can replace him in these predictable moments, but who also act in the unpredictable, uncontrolled moments within the social world.

Unlike the priest, the rezadores and benzedores have no formal education - although their knowledge is derived from the priest and passed down through oral tradition. It is the prayers they hear and memorize that form an important part of this *body of* knowledge. This transmission in the village of Itàunas is done within the same lineage, within the same family, as a family tradition, a *gift* from God, a vocation, as they reported in the interviews. There are three benzedeiras in the village and one rezador. The three benzedeiras interviewed are from the same family as the rezador, they are all related. The granddaughter of one of the benzedeiras, a young girl, has also shown a *gift*, and has already begun the art of benzimento, through her grandmother, albeit timidly, as observed during the fieldwork. See excerpt from interview with local benzedeira, 81 years old:

I have a granddaughter... she was growing up... now she's there... she's a year old... she's there with this business of this notary's office... and she intended to have it blessed...

And a girl came here to bless me... and I was eating.

Then she said: Grandma, I'm going to bless it! Then she picked the leaf and went to bless it...this girl and the girl got well! (Woman, 81 years old, interviewed in 2007)

For both rezadores and benzedores, knowledge of the right words and their correct order is of great importance; what differentiates them is the context, how they use the prayers and the functions they are assigned. Continuing, Zaluar (op.cit., p.113) adds that male pray-ers pray on major festive occasions, or whenever several people gather for novenas, benditos and ladainhas and their prayers have the effect of paying a promise to the saint or simply praising him. Their context is public. Benzedores, on the other hand, pray over the sick, over those afflicted by some ailment, where they know that their prayer is effective. Their context is private. Women can be pray-ers or blessers, but it is rare to find women leading public novenas or litanies. It is up to women to carry out the ceremonies in private places. Women benzedeiras have the exclusive right to help with childbirth, as midwives, to cure childbirth problems, as well as children, and in general every midwife is also a benzedeira (ZALUAR, 1983), but not every benzedeira is a midwife. In the village I met an 85-year-old midwife who no longer delivers babies because of her advanced age. Her knowledge has not been passed down through the generations, leaving no representative of the trade.

Figure 41: 82-year-old benzedeira and her daughter at the São Benedito and São Sebastiao festivities. Photo Xavier (2008).

Figure 42: girl (23 years old) granddaughter of the benzedeira (82 years old) and her son at the Ticumbi festivities.

Photo: Xavier (2008)

Figure 43: 74-year-old rezador with his grandson at home.

Photo: Xavier (2008)

Figure 44: 77-year-old benzedeira in her daily life. Photo: Xavier (2007)

Figure 45: 77-year-old benzedeira, one of the most respected in the village. Photo: Xavier (2007)

With regard to childbirth, there is a certain exception[100] in the village of Itaúnas, since a man (81 years old),

100 An exception in terms, since the supernatural 'entity' or 'spirit' that the healer receives is female: an old midwife called *Véia Andressa*.

the husband of a benzedeira, told me that he also "received" (incorporated) an "old midwife" (in his language) who worked in complicated childbirth situations, where the official midwife could no longer handle the job. He doesn't do it anymore either. In this case, only divine providence, through the *spirit of an old midwife called Andressa*, with her divine power, as reported in the interview, was able to relieve the parturient at this difficult time:

> You know what, ma'am? Look, lady, now we're off our feet! But when there was no doctor here, it was for everything, you see? Do you understand that it's for everything? For everything! For every problem, for everything..., for every problem the people had!!! You don't understand..., yet..., it was everything![emphasis] Are you out of the loop, lady? laughs.... [half mocking my lack of understanding] it was for everything I'm talking about. For all kinds of illnesses..., I was a woman who wanted to have a baby... it was everything! I did it! I had a midwife, Andressa! An old lady who *took* me..., who worked with me! (Man, 82, interview, 2007)

This account shows the power of faith and is, at the limit of understanding, an example of what Lévi-Strauss (1975, p. 216) calls "symbolic efficacy", where he describes a birth being conducted by a shaman. The same situation is repeated, where the intervention of the shaman is required, but rare, and is carried out in the absence of a successful birth, at the request of the midwife. Difficult childbirth would be a deviation, because childbirth is something natural. As a deviation, difficult childbirth requires the intervention of the supernatural, of protective spirits who can bring what is out of control, the physiological disturbance, back to its natural state of balance. See what the aforementioned author says:

> Difficult childbirth is explained as a detour, by the 'soul' of the womb, of all the other 'souls' of the different parts of the body. Once these are freed, the other can and must return to collaborate. Let us emphasize from the outset the precision with which indigenous ideology delineates the affective content of the physiological disturbance, as it may appear, in an unformulated way, to the consciousness of the patient. (LÉVI-STRAUSS, 1975, p. 216)

From another perspective, Lévi-Strauss says that healing consists of making intelligible, in terms that are affective and acceptable to the spirit, a given situation of pain that the body can no longer tolerate. For this author, it doesn't matter that the mythology of the shama doesn't correspond to an objective reality: the patient believes in it, and she is a member of a society that believes in it. That makes all the difference. The illness is resignified and finds an order in the cosmological order of the suffering social actor, in his being in the world. This idea is not far from Gadamer's experience of art (op.cit.), in fact it is in line with it, although Gadamer when he says that "being that can be understood is language" goes beyond linguistics. It is through this play of representations that symbolic power operates, bringing about transcendence and ultimately healing and well-being.

The male rezadores generally pray to the fields, animals and snakebites. In the village, I met a 74-year-old man with the designation of 'rezador', a cousin of the two other benzedeiras in the village. He said that he is also a benzedor, but his blessings are different from those of the benzedeiras. His specialty is *blessing fallen spines, open chests and astonishment*. In the interview, he said that he had been blessing since he was very young, since he had learned to bless by observing a benzedor who practiced this trade in the old village:

> Sometimes I'd ask someone to bless me and I'd pay attention, right? And I've blessed several people. And there are days when there are three or four people. And I'm black, but I'm not a pembeiro! [he says laughing a lot]. But thank God! Wherever I put

my hand with faith in God... God first, right? The person who believes! In the word of God, then...

There are people who pray and that's that! But I pray espinhela calda. Not that I'm any better than anyone else, but as I've learned, as I've been taught, I pray. I pray the espinhela first, then the open chest... I pray with a needle and a piece of cloth, with thread, right? [and he explains how it works] Cooking... On the day I finish praying, I say a padre nosso to a saint, or Saint George, or Our Lady of Desterro, or Our Lady of Aparecida! Any saint is valid. I take a cloth and boil it, saying the words. I bless when the person comes to see me. And each one has different words! (Man, 74, interviewed in 2007)

Like the local benzedeiras, this rezador is also knowledgeable about plants used for medicinal purposes, and the knowledge of these specialists is broader than local common sense, since it also covers plants used for magical purposes, as we'll see below. This *knowledge* is presented as an intimate relationship with the lived space, the society-nature relationship.

6.4 Diseases and their Remedies

The category of "illness" or malaise in Vila de Itaûnas is always in opposition to "health" or well-being. Illness is thought of as something that comes from outside and for some reason settles in the person's body, causing an imbalance. This imbalance affects the family and the community in degrees ranging from mild to very severe. Healing comes from re-establishing the lost balance (XAVIER, 2004).

For Foster (1976, apud Greco Rodrigues, 2001, p.132) there are two types of classification in what he calls "popular medicine"[101] . This differentiation is based on the explanation that each cultural group gives for the origin of their illnesses. According to the prevailing interpretation, he called the classifications personalistic and naturalistic. The personalistic types refer to illness when thought of as having its origin in the mystical-religious universe where the individual is a victim (for example, the evil eye, the broken heart). The naturalistic type indicates that the illness is understood as having its origin in natural forces (such as whooping cough, measles, chicken pox, among others). In which the individual is not a victim, but an agent of their illness (as related to something they ate or where they were). However, it is possible to agree with Greco Rodrigues (ibid., p.133) when he says that these systems are not pure, and there may even be an imbrication of them. This imbrication occurs mainly in the therapy undertaken by social actors who use magical-religious rituals for illnesses considered naturalistic, such as among the residents of Vila de Itaûnas.

In the art of healing, the specialists in the village of Itaûnas use what they call home-made or bush remedies, made from plants used for medicinal purposes, as well as prayers, blessings and plants with magical powers, which are harvested at a certain time in the calendar, including the lunar calendar. These elements are amalgamated through healing rituals, or magical rituals, with the use of sacred words and chants that turn them, within a gnosiological logic, into agents for promoting healing. Although the official healers are the specialists, in general the women and men of the village of Itaûnas have a collective repertoire, diffusely taught orally, and diffusely learned; but which operates through an efficient pedagogy, which ensures that the knowledge about plants for medicinal purposes, and other elements that promote the relief or cure of their ailments, circulates and is reproduced through the generations. The repertoire of plants used for medicinal

101 Although these are not considered popular classes, as I have already said, I decided to put this classification because I understand that it is one way of understanding and there are others.

purposes is very well described in the monograph by Uzelim (2005), which can be consulted, and the booklet produced by Cia de Oficios da Terra is in the appendix of this book. In this case, I'm only going to focus on the interviews

Greco Rodrigues (2001, p. 132) says that like any other field of knowledge, popular medicine, which I refer to here as the art of healing, has its professionals who have a particularly large body of knowledge or possess a particular *virtue*, a *gift*, and work as benzedeiras, rezadores, raizeiros and midwives or curandeiros. However, this is a universe where knowledge is not exclusively appropriated by these specialists, and so it is shared by the whole group, to varying degrees, and according to individual interests and needs. This shared knowledge makes the woman, the mother of the family, the main therapeutic agent, or the first aid in the event of illness or discomfort. The woman in this society has knowledge, and through this she makes the first diagnosis, as a first resort, in any illness that appears in her family, until they can access a medical specialist, or a benzedeira, pray-er or midwife, if necessary.

The men in the village show that they also have therapeutic knowledge, but for more practical and specific purposes. They use medicinal plants in *pingatherapy*[102] , in other words, they add medicinal plants to cachaça to 'curtir' it (in the local language) and thus try to extract its therapeutic purposes or principles. The plants used for this purpose are notorious for their aphrodisiac powers, for closing up the body, for giving it physical resistance, for thinning the blood, for warming up the body, as well as for curing colds, etc. The best known are: cinnamon stick, ginger, clove vine, milome vine. Some drinks are made from a mixture of these plants, sweetened with honey, such as the popular 'ximboquinha' and 'cipó- cravo', which are widely used by forró dancers and sold as a stimulant to tourists who enjoy them very much.

The best drinks, the most traditional, are sold in the small bars (where the owner makes the drink himself), those most frequented by the local population, by fishermen, and which offer no attraction to tourists. The plants used in these drinks are generally considered to be 'hot' plants, linked to the masculine character in Hippocratic medicine, which we'll see below.

The plants used by women, on the other hand, are generally for first aid and those illnesses that they take care of at home and are considered mild, such as: children's illnesses such as physiological disorders caused by tooth eruption (fever, vomiting, convulsion, nervousness), worms, colds and flu; and also illnesses for which there are already vaccines, "but which still give, only lighter" (in local speech), according to the interlocutors, these are: measles, chickenpox, mumps, rubella, whooping cough.

There's chickenpox, there's variola. Chickenpox, you don't have to bathe it when you're running a fever, do you? To make it thick... then it gets all those little balls... just like measles. Measles... if it's measles... you have to wait and see if it's measles that's going to break out. If it starts, you leave the child dirty and then give it a bath. If you give it a bath, it will pick it up. Don't bathe it... wait for it to sprout! Take elderberry tea. (Woman, 51, interview 2007)

In the village of Itaúnas, I found four categories of remedies, as interpreted in this work: 1 - Pingatherapy - a stimulant, but it can also be used as a remedy to restore certain physiological functions, increase virility, cure

102 Neologism used by Araùjo (1961), in Medicina Rùstica Brasileira.

flu, increase appetite, close the body against the evils that come from outside, warm the cold body, thin the blood. These plants and others can also be tanned in wine, adding Biotònico Fontoura, a popular pharmaceutical formulation that, when mixed with certain plants, has the effect of 'fortifying' the blood. Some are known as garrafadas and are widely used in rural areas.

2 - Home remedies, or bush remedies - are used for known illnesses that can be treated by a doctor, but are usually resolved at home, as they are considered 'mild'. Plants are used for medicinal purposes and/or animal excrement, plant liniments, roots, bark, resins, with the aim of restoring health or well-being. Their efficacy is based on local empirical knowledge, whether from experts in the healing art of benzedeiras, rezadors and midwives, or even on the common sense that tradition has made official through the generations.

3- Blessings, prayers and baths - this is a remedy for illnesses that find an explanation within the local cosmovision. In general, they don't go to the doctor, because "it doesn't work, or it doesn't do any good". They are considered to be illnesses, or disorders, where the etiology, treatment and cure are supported by a pray-er or a benzedeira. The illnesses treated by benzedeiras are: espante, quebrante, espinhela caida, mal olhado, perto aberto, cobreiro. Midwives are also blessers and use their prayers and magic words during childbirth, with particular devotion to Our Lady of Sorrows, Our Lady of Good Delivery, Santa Barbara, St. Bartholomew[103] , Our Lady of Montserrat.

4- Pharmacy remedies - are those prescribed by a doctor. Sometimes these remedies don't achieve the desired goal. In this case, they say that "they don't always get it right with the medicine", or the medicine doesn't solve the problem and so they also use teas, or home remedies, at the same time.

Chart 2- Medicinal plants and other remedies and their therapeutic purposes.

Bush or home remedies	Therapeutic purposes
Marcela	Soothing for tooth rash, stomach ache. Take a bath to calm down and give the tea to drink.
Sapote root	Cha helps 'break in' the child's first teeth.
Taririquinha	Tea is used for colds and flu, sweetened with honey.
Elderberry	Cha is used for measles fever.
Carambola	Tea is used to regulate high blood pressure.
Lavender	The bath and tea are used for menstrual cramps, headaches, body aches and general malaise. It is also used for unloading baths, baths with scented plants to bring good luck. It is one of the plants most mentioned in the interviews and found in

103 It must be Saint Bartholomew, the former patron saint of the town of Itaùnas.

	many backyards.
Monkey cane	Tea is used for long coughs or whooping cough.
Holy grass	The juice of the plant is mixed with other plants for the same vermifuge purpose.
Boleira	The juice of the plant is mixed with other plants for the same vermifuge purpose.
Mint	The juice of the plant is mixed with other plants for the same vermifuge purpose.
Parsley	Tea is used for inflammation of the uterus.
Holy grass	Tea is used for headaches, as a calming agent to soothe 'worms' and to calm children and adults in general.
Rue	The tea leaves are used to help expel the fetus during childbirth.
Field jasmine or field flower (dog droppings)	The white part of the excrement is collected, placed in a cloth and tea is made to soothe measles.
Saffron	Used to apply around the eyes of those affected by measles.
Cambucà	The bud as tea is used for whooping cough.
Carnation	Tea is used to increase labor pain and speed up the delivery process.
White onion	Used to 'cool' the belly of the woman giving birth.
Arnica	Boiled arnica leaves for sitz baths, for healing after childbirth.
Erva santa, or Santa Maria	It was used in sitz baths for healing after childbirth.
Cashew shells	I used to boil the cashew bark and use it in a sitz bath for healing after childbirth.
Milome vine	Closes the body against evil, improves male functions, good for stomach problems.
Melao de Sao Caetano	Vermifuge
*Neovalgina (named after the plant)	Headaches and fever
*Doril (named after the plant)	Headaches and fever

Araçà	Belly pain
Cambucà	Belly pain
Castor oil	To heal a newborn's navel

*[104]

6.5 Blessings and their rituals:

In the village of Itaûnas, I met three benzedeiras and a rezador, but there are others I didn't have the opportunity to interview[105] . The oldest interlocutors in the village told me that there were many benzedeiras in the village, and around three midwives. In the old village Dona Gertrudes (no information), in the new village Dona Conceiçao (deceased) and Dona Tidù (85 years old). Those who perform the art of healing - midwives, benzedeiras and rezador - don't receive any money for their services, since for them this is a divine *gift* and they do it in return for their devotion to the saint. As I've already said, the Vila's healing officers come from a lineage, a family that has this *gift* or *vocation*. They learned it from their mothers, grandmothers or close relatives like an aunt. The rezador, on the other hand, had *the gift* very early on, but went to learn from another rezador.

These specialists in the art of healing carry out their sacred work in their homes, with their oratories, in front of their saints and through their magical words. Every ailment and illness has a type of prayer, a saint who acts on a particular organ of the body; the benzedeiras and rezadores treat both bodily ailments and ailments affected by spells and magic, such as the evil eye.

For the ritual, the specialist usually uses three branches of rue or three branches of tarariquinha, or another plant of her choice. While praying, she passes the branches of the plant over the devotee's body with one hand. At the end, if the twigs have wilted, it's a sign that the body was "loaded" or "heavy".

6.5.1 Evil eye:

It's something that comes from outside, in the relationship between beings and the world, in the world.

"Good man, bad woman, against breakage and the evil eye! Our Lady who takes away this evil eye, this broken eye, this evil eye, this evil eye; that is in the flesh, that is in the bone, that is in the blood, that is in the nerve, *is from so-and-so (say the person's name)*. And Jesus takes it away and throws it into the waves of the sacred sea, where the ox doesn't scream or the rooster sings, with the words of God and the Virgin Mary." This is how we bless! (Woman, 44, interviewed in 2007)

"Lord Jesus, what do you bless? Then Jesus says: I bless you from all evil: breakage, evil eye, ailment, right? And he throws it into the waves of the sacred sea, where the ox doesn't scream and the rooster doesn't sing, with the words of God and the Virgin Mary!" (Woman, 44, interviewed in 2007)

"God is the Sun, God is the Moon, God is the father of kindness, God is the merciful father, who takes away all the evil, all the evil, all the evil eyes and throws them into the waves of the sacred sea, right? Where no ox roars, no rooster sings, with the

104 See Araùjo (2002) for more on this phenomenon of giving the fantasy name of a drug to a plant used for medicinal purposes.

105 Conf. especially the careful work that Uzelin (2005) carried out with these healing officers in the village of Itaûnas.

words of God and the Virgin Mary!" (Woman, 44, interviewed in 2007)

According to this benzedeira, it's not enough to bless, you have to *lock the body, front and back.* In this case, she uses the words: *Mary and Joseph, Mary and Joseph*.

6.5.2 Looked at, amazed and broken:

It is also related to the relationships between beings, and is something that comes from outside.

"Good man, bad woman, old lie, enchantment and gaze. May Our Lady take away these eyes and carry them to the waves of the sacred sea!" (Woman, 82 years old, interview 2007)

"God is the sun, God is the moon. God is the three persons of the holy trinity, get out of here, look, scare, break, and four ailments. All the evil you have in your body, spells, curses, go to the waves of the sacred sea. May Santa Barbara come and cleanse you!" (Woman, 82 years old, interview 2007).

6.5.3 Distressed:

It's when a person makes an awkward movement and there's a twist.

6.5.4 Open meat:

It's when you feel pain in some part of your body, but it's not a twist, it's the flesh that has opened.

"What is sewing? **A stitch of open flesh, a twisted nerve, a** cracked vein, that's how I sew. With the words of God and the Virgin Mary! If we just want to say it like that, if it's **just distroncadura**, we should say it like this: what do I sew? I sew distroncadura, with the words of God and the Virgin Mary. Right here I sew, with the words of God and the Virgin Mary!" (Woman, 44, interviewed in 2007) [emphasis added].

6.5.5 Froggy:

Thrush is a disease that occurs in the mouths of young children.

"Frog, little frog, jump here, jump there! Just as Our Lady left us the host at our wedding, get out of here, little frog, with the words of God and the Virgin Mary!" (Woman, 82 years old, interviewed in 2007 - first prayer she performed when she was still a little girl and already did frog blessings)

6.5.6 Copper:

A rash with vesicles, similar to herpes simplex, is called mumps.

"Frog, toad, snake, snake. Every animal in a nation, all I ask is that it doesn't bend over and touch its tail with its head. Santa Elisa had three children, (incomplete?) the one on the hill, the (?) and the wild copperhead. Saint Elisa asked Our Lady, what would you do? Cut off the head, cut off the tail, with the words I will heal." (Woman, 82, interviewed in 2007)

6.6 Unloading bath:

They use baths with scented plants (local language) and unloading baths when the body feels heavy and tired for no apparent reason.

"You put cold water in a basin that you can take a bath in! You go to the coffee tree... take 15 leaves... 15 green coffee leaves [emphasis added]. Go into the water... tear it all up... take a good look!!! huh?... tear it all up into the water. Then you go and put a bit of perfume... into this water.

His wife says: the best-smelling perfume!

Then you add three pinches of salt to the water. Then you mix... you mix everything... with those little torn leaves, the salt, the perfume... with enough water for you to take a bath. When you've finished, you stand over the water and say a prayer... to whichever saint you have the most faith in! To the saint... and bathe from head to toe!

Say a prayer over the water like this... (making a cross) and ask for what you need most!!! Ask for what you need from your feet.

But don't throw that water away for nothing... you go to a little place where no one can walk over it and pour it in like this....

This is the unloading bath.

The salt is to baptize the bath! Don't you know you baptize with salt? If you have holy water, add a little! But holy water is only used in church... You pray to the saint who has greater faith than the others... well, then..." (83-year-old man, interviewed in 2007)

"From time to time we have to 'take care' [emphasis added], right? ! We have to take a bath like this... to unload... to be able to close a body!!!

Oh my God! ! The leaves, eh? [pensive...] Basil... there's some...

Basil, rosemary... what else? [struggling to remember] Tiririquim, rue. Tiririquim, don't you know it?

[the more fragrant the herbs... the better! All that. The more the bush smells the better!

Cidreira, it's all good... then in the bath, you go... puts a.... and doesn't cook, eh? It's rubbed, it's rubbed in the water... and when it's finished, to get the madness out of the leaf... when it's finished. Then add a little coarse salt!

Then he takes a bath... with his head on and everything. That's an unloading bath!!!

[head and everything?] head and everything! That's an unloading bath!" (Woman, 82, interview, 2007)

6.7 Faith that heals.

The accounts and interviews demonstrate faith as symbolic effectiveness. Both the officials in the art of healing and the clients/devotees all responded that what really *heals is faith, and if there is no faith, it's no use.* In this case, faith works like the will, the willingness to *play the game* (a game of representations). Gadamer (1985, p. 39) says about the *movement in the game*: such movement in the game means, at the same time, that playing always requires the one who is going to play along. (...) When I really "go along" it is nothing other than participation, inner participation in this movement that repeats itself - it is the hermeneutic identity of the game - the being-ai (dasein) being. It is through this "going together" in the game of representation, as in the staging of a play [106] , that transcendence, healing, is achieved.

"The other day when eia left, eia came here with a little girl for me to bless. Her granddaughter... her daughter's. Chubby, chubby... tiny.

It was bad... I couldn't sleep...!!! I didn't want to sleep...

But they've gone and I don't know any more about it, do I?

But it got better! With faith in God, it got better!" (Woman, 82 years old, interviewed in 2007) "Intào...it was this one, the Holy Mary. We used to take more of this one. That was our medicine. And the prayers...the trust we had in God, right? There were

106 Theater actors and spectators in a game of representations. Theater was born out of this context of religious worship as representation (GADAMER, op. cit., p. 40).

prayers to St. Bartholomew... É. He's the women's lawyer!" (Woman, 44 years old, interview, 2007)

6.8 Causes of pregnancy and childbirth.

Pregnancy is also a state of passage, a liminal state. These rites accompany every change of place, state, social position, and age. *They operate by indicating the contrast between "state" and "transition", employing "state", including all its other terms* (TURNER, 1974, p. 116). All rites of passage are characterized by three phases: separation, margin (or *limem*, meaning *threshold* in Latin) and aggregation. In this liminal state, beings have no *status*, no property, no insignia, no position, no kinship, in short, nothing that can distinguish them from their neophyte colleagues or those in the process of initiation (TURNER, op.cit., p. 117-118). In this situation or process there is a very close relationship between society and nature, and everything is related, which is why it is a phase full of mysteries, desires, interdicts, a non-position in a waiting phase. These *cases* cited by the informants demonstrate the beliefs in the relationship between microcosm and macrocosm, and the relationship between beings and things, the intimate relationship between society and nature that is possible in the state of liminality.

"It's very dangerous. You can't put a key inside your breast. A leaf... it comes out, the child comes out straight, the child comes out."

"I couldn't, I couldn't pass over things, I couldn't pass over things that were tied up. Because that's the example, right? My cousin's wife even died, she... made the pamonha marrada, you know. You can't tie it up with a little rope, right?

She passed... she was pregnant at first, she passed over, my daughter, the boy came out just right. Bracinho, gualzinho a pamonha marrada". (Woman, 44 years old, interview 2007)

"At the beginning of pregnancy. You can't, if there's a big train on TV, you can't say it at first. That it comes out, I have a suspicion that it comes out... Just like that. If you notice it, it comes out!

[But it's noticing, how? I ask] I think, if you criticize, if you laugh... right?

Intào! It's dangerous, why did that girl happen, oh! She was laughed at. She was born right with the rat. She died. Right in the mother's hand, right. It's dangerous! You can't be forty like that, pregnant at first, you have to keep everything in line *(laughs)*. It's milindroso. Ééé...milindroso. You can't!" (Woman, 44 years old, interview 2007)

6.9 Safeguarding and its prescriptions given by midwives.

For the women of Vila de Itaûnas, confinement lasts 30 days. It is also a phase of transition and liminality, both for the mother and for the baby, who does not yet have the status of a person within the social group. It is believed that a woman's postpartum body is open, both physically and spiritually, and that is why she needs special care, divine protection, the protection of the other women in her family and the solidarity and complicity of her partner. In this case, the woman's body must be protected from the cold, wind and evil spirits for 30 to 40 days. This belief that a woman's body is open is culturally reproduced throughout Brazil, especially in rural areas. In the research on Saco do Mamangua, the postpartum restrictions and recommendations were very similar to those found in the village of Itaúnas, as also found by Motta-Maués (1993) in his work on the communities of Itapua (an Amazonian fishing community).

"From the day of childbirth, the **woman's "grave" is open**, and when the forty days are up, it closes and the woman is free. It is also on this day that the woman's body "closes", after having been "open" since pregnancy, hence the need for her near immobility and seclusion. For all these reasons, any carelessness on her part at that time could lead, they believe, to serious inconvenience and even death. Restraint doesn't really end at the end of forty days after childbirth because, as far as dietary prescriptions and certain activities are concerned, this period extends to a year after the birth of the child." Motta-Maués (ibid., p. 144)

During this period, her diet had to be special; she was the one who would be fed and cared for first. The midwives were the ones who gave her the regime, and they also said their blessings and prayers.

Figure 46 : Dona Tidù, an old midwife, and a representative of the Ticumbi during the festivities. Photo: Xavier (2008)

Figure 47: Dona Tidù, an old midwife from the village of Itaûnas, and part of her family. Photo: Xavier (2008)

Today, women go straight to the hospital and rarely deliver at home, since there are no longer any official midwives in the village, which represents a great loss for the local culture. Some women obey the restrictions and their prohibitions, others no longer, but some prohibitions, especially food ones, are still used, including foods that are considered heavy. The restrictions and prescriptions of this period, called resguardo, have the function of protecting the life and health of the woman during this passage, since her "body is open" and receptive to illness. If the passage from the 1st to the 4th period (40th day) is successful, the subsequent periods have restrictions and prescriptions that are considered to be increasingly milder, since by the 40th day the body has closed down and there is no longer any imminent *"risk of life"* (risk of death). However, dietary restrictions on fish and 'heavy' or 'reamy' foods persist until the fourth month after childbirth. The first period is quite expressive, revealing the care taken to ensure good healing, and sitz baths with leaves and bark of plants with an astringent effect are recommended to close the body and heal the perineum, including: cashew bark tea, arnica leaves, yerba santa, St. Mary's wort, lavender and artemisia (XAVIER, 2004). During this period, cold is forbidden, a concept also used in Hippocratic medicine, of the balance between the 4 elements (fire, air, water and earth), the seasons with their climatic states: hot, cold, dry and wet. The uterus is moist and hot, the blood is hot, the body is hot and open[107] , and cannot mix with cold things. See table 2, with the prescriptions given by the old midwives of Vila de Itaûnas.

Chart 3 - Prescriptions of the 'old guard' given by the local midwives.

	Diet		Non-food restrictions	
	Permitted foods and/or recommended s	**Foods banned**	**Permitted and/or recommended behavior**	**Prohibited behavior**
1st period: Day 1 to Day 3	Scalded farm chicken, biju with coconut, coffee with cookies.	All the others, including rice and beans.	First child: Bathe only with lukewarm water; take a sitz bath with lukewarm water and tea made from cashew nuts, arnica leaves, holy herb, santa-maria. Lavender, artemisia. Breastfeed the newborn, only get up to go to the bathroom. The first period is the hardest.	The first child: Getting cold on your feet, head and ears, letting the wind blow in your ear, having sex. Lifting weights, sweeping the house, blowing on the wood stove. Cold or iced liquids. Doing household chores
2nd period: Day 4 to day 8 or day 10	In addition to the above diet, baked, fried or moquequinha fish	Fish considered 'heavy' (leather or sandpaper): calafate, peroà, baiacu, arraia,	Breastfeeding. Take sitz baths if necessary. You can walk around the house, receive a visitor, do some	Catching cold, doing laundry, heavy lifting. Having sex

107 Check out table 2 based on the *Corpus hippocraticum.*

	should be added. Black pork scalded meat, you can also	caçao, caranguejo, siri, ostra, camarao. Heavy game meat: capybara, catitù.	light work in the kitchen.	
	to be roasted. Light' (scaled) fish: white catfish, jundiarte (river catfish), roncador, hake and bearded catfish. Rice and black beans (only seasoned with garlic and salt) and a small piece of rib). Light game meat: paca	You can't eat eggs, they're hot. No food coloring. Drink alcohol. Foods considered "heavy" or with "reams" such as duck, turkey, pork other than black, beef. Heavy fruits: watermelon, pineapple, jackfruit, which leave a woman feeling tired.		
3rd period: 8th day to 30th day or 40th day	The diet above	All of the above	She can now do some laundry and household chores that aren't too heavy so as not to "strain" the uterus. Breastfeeding	Have sexual relations. Avoid heavy work
4th period: from the 30th or 40th day onwards.	The diet above, with the banned foods being prohibited.	Avoid 'heavy' foods until the 4th month, including shrimp, duck, capybara, catitu, pork that isn't black.	Breastfeeding for up to one year is recommended. Household chores. Sexual relations.	Avoid heavy lifting until the 4th month.

6.9.1 *Heavy* or *rowdy* or *loaded* food.

For local actors in the village of Itaúnas, food is considered *light, heavy* or *bland*. Heavy or bland foods are avoided during periods of convalescence for children, the elderly and adults, in the postpartum period or in confinement, as well as when breastfeeding. According to Xavier (2004, p. 144) *reima* comes from the word *rheuma* (a pathology), evoking the idea of a catarrhal or watery flow of humor. In the 16th century, the word *rreyma* was used, from the Latin rheuma, derived from the Greek *rheûma*; the latter also represents acts, actions and behaviors (CUNHA, 1998, p. 683). According to Ferreira (1999, p.1734), reima is a corruption of reuma, from which comes reumatismo, a word always linked to inflammation. "Reima and reuma" can also represent mal-gênio (mood) and rabugem (attitude). Foods considered *heavy or rheumy* correspond to foods with "ream or reima".

But Greco Rodrigues (2001, p.140) says that the reima or ream principle is based on the humoral principle of Hippocrates, a Greek physician (460-377 BC). Hippocrates is considered the father of medicine for codifying all the medical knowledge available at the time and organizing it in a repertory called the *Corpus hippocraticum*[108] . According to this *Corpus*, reima would be the flow of the humors and reimoso would be that food or attitude *capable of disturbing that flow*. So the potential for reima or *remoso* is related to *organisms* in disharmony, in the most fragile, delicate moments and not exactly to food: *food is reimoso or remoso* to. Food that is reminiscent and heavy for one person may not be for another, depending on the context the person is in.

But where does this knowledge come from?

This medicine came to Brazil with the Jesuits in the 6th century, and Rodrigues (ibid.) reports that although Brazilian folk medicine, the art of healing, has received contributions from various cultural groups of blacks and indigenous people, it is essentially a knowledge that came with the Portuguese colonizer, and resembles the medicine practiced in Europe at that time. Specifically, the Jesuits were the first to exercise this function in colonial Brazil. Here is a brief explanation of the subject.

The origin of this medicine is in ancient Greece, and we find its system related to the knowledge found in the *Corpus hippocraticum*, as already mentioned. Within the principles of this Hippocratic medicine, the notion of balance is paramount, and forms the basis of the explanation of illnesses in our healing art of the healing officers in Brazil. According to this theory, it is the humors that, through the balance of their qualities, should keep the organism healthy, and disease is the disruption of this balance. Disease sets in when one of the qualities of the human body gains predominance over the others through the action of an internal or external agent. This condition varies from one person to another, in different contexts, and according to each individual's nature: hot, cold, dry or wet. Ways of explaining the construction of the universe through pairs of opposites, in natural oppositions such as life and death, right and left, light and darkness, hot and humid, raw and cooked, are found in almost all the symbolic systems of many other peoples, not just the Greeks. Hippocrates (460-377 BC) was certainly not the only one to emphasize the importance of balancing the forces in human bodies, and we know that Indian Ayurvedic medicine and Chinese medicine, among others, also do so[109] . In this cosmology, the symbolic categories are correlated within a gnosiological system and present this relationship between man and nature. Rodrigues (ibid. p.135) says that the classifications are based on the concept that people, diseases, medicines, food and most natural objects *have nature.* Within this system, medical practice consisted of understanding the *nature of the patient*, determining the *nature of the illness* and restoring the fundamental harmony that had been disrupted.

108 Cairus (2005, p. 25) in The *Corpus Hippocraticum*. See also table 3 on p. 132.

109 Samuel Hahnemann's homeopathy uses the knowledge of Hippocratic medicine.

Although the basis of the medicine practiced on the Iberian Peninsula in the 16th century was Greek medicine, it should not be forgotten that this knowledge was the result of multiple influences: Celtic beliefs, the principles of Roman and Greek medicine, the use of Christian prayers and blessings and Moorish beliefs. The transfer of this knowledge to the new world took place through formal and informal mechanisms, as the State and the Church developed strategies to guide colonial policy, and at the same time contact with the native peoples adapted Iberian customs to the new reality. Imagine adding to this knowledge the knowledge and beliefs of the Africans who were brought here as slaves. In this sense, we can see that there is no single strand of knowledge in the Brazilian art of healing, but some stand out according to how they were disseminated in local traditions. This is why they vary throughout Brazil, but they retain the main aspects of Hippocratic medicine.

Food with reimosal potential (or remoso, in local parlance) is generally related to occasions where organic flows (two organisms coming into contact), of a normal or pathological nature, appear: menstruation, puerperium, intestinal disorders, wounds or expectoration. On all these occasions, when the internal humors are exposed, the body is more *vulnerable* (in liminal states, convalescence, illness) and reminiscent food has the potential to disrupt this flow. Like the humors, reima is associated with the problems to which these humors are related, especially blood and its quality of being hot. This creates a tendency (very common, as seen in BRANDÂO, 1981) to define reimoso or *remoso* with something that is hot. Greco Rodrigues (ibid., p.141) also informs us that some foods considered "strong" have a tendency to be reimoso, because their potential aggressive force disturbs the organism. In the village of Itàunas, foods that are considered heavy tend to be *rowdy*.

In the village of Itaûnas, local actors classify pork (other than black pork - native pork) and duck, fatty beef and some game as *heavy* or having the potential to *ream* or *paddle*. This means that it interferes with the humoral fluids, causing imbalance. In the village of Itaûnas, women in a *state of passage,* such as "resguardo", should avoid these foods, as they can trigger the *reima* process, disturbing the humors and leading to serious imbalances. It can also mean that food with reima potential, in people with open wounds, children convalescing from hot illnesses (where there is a fever process) such as measles, chickenpox, whooping cough, rubella, and mumps, can trigger inflammation and/or increase inflammation, fever, because the wound or illness can get worse (the tension between *the hypocritical natures*, *excess and lack*). In a ranking from highest to lowest in terms of weight: pork (other than black) > duck > leather fish, or sandpaper > game (catitu, capybara) > egg > beef.

> It [mumps] comes out behind the ear with a little bump... and the fever comes first. Then it starts coming out... some children get two at the same time. Then it goes down... everything gets swollen... then you only have safflower to bathe with. You can't eat eggs, because eggs are hot... You can't eat *loaded* food like that..........(Woman, 51, interviewed in 2007)

Heavy or *unhealthy* foods, which provoke or interfere with the normal flow of the humors and generate or aggravate pathological states, are generally of animal origin, although some vegetables are on this list (BRANDÂO, 1981, p. 95-152), such as rice, which is not recommended in the first days of confinement, beans (except black beans), pineapple, jackfruit and watermelon. Duck is forbidden, according to the informants, making it unsuitable for certain periods of life and certain people. The free-range chicken, on the other hand, is appreciated and recommended in the border states.

Thus, we present the Hippocratic medicine as the guiding thread of this gnosiology of the art of official healing, of popular Catholicism, in the village of Itàunas, interwoven with the many other knowledges that made up this *corpus of* knowledge. These knowledges, practices and arts are not isolated, as they may seem, but are continually articulated with biomedical knowledge, with the aim of solving life's practical problems, but always in the search for balance, well-being, joy, in the life of relationships. See table below:

Table 4 - Representation of the society-nature relationship according to the *Corpus hippocraticum.*

The qualities	The four elements	The four parts of the world	The four winds	The four parts of the year	The four humors	The four ages
Hot and humid	Air	Noon	South	Spring	Blood	Childhood
Hot and dry	Fire	East	East	Summer	Cholera	Youth
Cold and wet	Water	West	West	Autumn	Phlegm	Old age
Cold and dry	Terra	Northern	North	Winter	Melancholy	Decrepitude

(CORTEZ, s/d, p. 23, apud GRECO RODRIGUES, ibid., p. 143)

In these ways of thinking and acting in the health/disease/cure process, we find an intimate relationship between society and nature, microcosm and macrocosm (GRECO RODRIGUES, ibid., p.143), which are presented in the lived space, as strategies and tactics for spatializing the world, a cosmology where these forms cohere, invent and reinvent a world. All in all, it is a symbolic capital and true intangible heritage[110] of the community of Vila de Itaùnas.

110 See Xavier (2004, 2005, 2006) on this *body of knowledge* as intangible heritage.

7. Reflections: towards a conclusion.

We've finally reached the *momentum of* this work with part of our concerns resolved, and others not so much. In these terms, this isn't exactly a conclusion, but a reflection that we intend to extend beyond this book.

We have seen that the aim of this book was to understand how the practices and knowledge of the officials in the art of healing in the village of Itaùnas produce and reproduce space in their *moments of being in the world*: spatiality, territoriality, constituting a place, and in an identity process, inaugurating a territory. So here are some final thoughts and considerations.

Within the limits of the interpretation of this ethnogeographical work, we consider that the actors of the Vila de Itaùnas are driven by devotion and religious faith, but not only, as we have seen represented in the association of popular Catholicism of the Institution of Ticumbi[111] . In the arrangement of this institution we find the officials (specialists) of the healing art as a sub-institution of the Ticumbi. Regarding Ticumbi, it doesn't seem to be exactly a Brotherhood of popular Catholicism, but it is very similar, especially when it comes to the orientation of lay agents, linked to a particular saint.

In general, the brotherhoods are made up of groups of lay devotees who organize themselves as private associations, not ecclesiastical ones, and whose purpose is to maintain a cult or devotion. They have been the greatest resistance to the clergy's reform process since the 19th century. Although they are dependent on the clergy for their devotional rituals, they maintain legal and economic autonomy from the clerical regime. The Brotherhoods[112] own the shrines and economically exploit the events.

In the case of Ticumbi, this is not the case, as the institution does not own the parish church of Sao Sebastiao, nor the church of Sao Benedito (which belongs to a local actor, who keeps it out of devotion), and it is transformed into a gathering of folklore groups, including an association and headquarters. In these terms, we argue that the folklore category is a strategy and tactic, a formulation of symbolic and territorial resistance to its traditions and customs.

Lima (2007) points to the territorial and symbolic difficulties faced by communities of Afro-descendants in a region called Mumbuca, in the north of Minas Gerais. As in Mumbuca, a large part of the population of the village of Itaûnas is a remnant of the Angelim[113] , and has strong ethnic and cultural traits of Afro-descendants. According to the interlocutors' accounts, this Angelim community has long suffered from threats to their territory, as well as historical prejudice against their religiosity, including reports of violence by the local police. This state power has threatened the meetings and games of congo, jongo[114] or capoeira[115] with persecution and imprisonment of devotees. The serious funding problem in the north of ES, in relation to quilombola territories, is a fact and has been evidenced over the last few years, being reported in various media.

---

111 Conf. with the Brotherhood of the Good Death, presented by Corrêa (2004).
112 See Steil (2001) and Corrêa (2004).
113 In the process of having its Afro-descendants recognized as "aquilombados" territory. Locality belonging to the district of Itaûnas.
114 As a report by Rogério Medeiros (2008) shows us, in a meeting with a lady from the rural area of Conceiçao da Barra, who had the memory of the only female Ticumbi group, Mrs. Erotildes de Oliveira, 84 years old.
115 Medeiros, 2008.

In this sense, the religious manifestations[116] of Afro-descendants have always been seen as folkloric and thus tolerated, but the conflict has always oscillated between explicit, as mentioned above, and implicit. We can infer that as a *folk*, the institution does not challenge or threaten the local powers organized as institutions legitimized by the State: the Church (clergy), the PEI (IEMA), the Corrégo Grande Biological Reserve (Rebio) (Instituto Chico Mendes, formerly IBAMA), the Conceiçao da Barra Municipal Government.

This traditional religious association makes up the arrangement of life and the calendar of the social actors of the village of Itaûnas, with its periodic rehearsals, organization of raffles, work on making clothes, ornaments, novenas, and general planning for the climax of the festival on January 18, 19 and 20. The tributes to the patron saint Sao Sebastiao, and the saint of devotion Sao Benedito, are presented as a great liminal ritual, and through this complex ritual of *Festa, folia or festejo* reinaugurates a territory, in the incarnation of an *ethos*; a place as its own, expressing *a sui generis* spatiality.

This spatiality, which at a certain *point* becomes territoriality, is presented in the Vila as a powerful geographic strategy, with the aim of influencing and controlling people, phenomena and relationships, which by delimiting an area as a territory reveals a lifestyle, as well as a cultural identity. In these terms, we seek to explore the various moments when these strategies show themselves more and/or less, but always present as an action by their agents.

For these reasons, we understand that this Ticumbi institution is not simply a *folk* or devotional movement, as it shows its strong political face revealed in the strategies, tactics and discourses that underlie the event, the one that appears on the surface, the revealed.

We realize that the residents of Vila de Itaùnas, especially the older ones, keep a historical reference, as a collective memory of the old Vila, the one that existed on the other side of the river and close to the sea. In the second chapter, we try to show this time/space of the place, bringing the reader up to date in order to give a general idea of the 'place' we are talking about.

In this work we are not looking for what is explicit, but for what was not revealed at first, what was in the relationships woven into the daily life of the social actors in Vila de Itaùnas. This led us to work with micro-geographies, those that cannot be understood at first glance, but can be grasped through living with symbols, meanings and signifiers. In this way, many moments of being in the world were delineated, which is why we understood that there was a need to explore the concepts (in the third chapter) with which we worked. We know that this can be uncomfortable, but imagine this book as scenes from a film[117] where at various times and from different angles, we capture a scale, a color, a detail, where a scene is presented as a whole.

These scenes are in the fourth chapter, where we explore and discuss the various moments of the conniving landscape in Vila. This conniving landscape was explored to the limit of its possibilities, given the difficulties inherent in the process of representing it in a work of art. We recognize that the ritual complex could have been more detailed, but the ethnographic *momentum* only occurred when I was *already* in the final stages of

116 They are part of an identity and territorial strategy, as we have already seen.
117 That's exactly how she ended up in this author's head.

fieldwork, and there was no availability to return to the field. Still on this chapter, we tried to emphasize as much as possible the movements of the conniving landscape, a non-static, more fluid landscape, which is the protagonist of the Vila's soul. This soul is in the life that everyday life reveals, because it is woven into the symbolic and objective exchanges that take place through the intersubjective interactions between local actors and their world.

These exchanges keep the community in a state of belonging, of community. But what are these exchanges? Relationships in a Vila are always close, marked by personalities. On visits to the curing officers or sitting in the best vantage point in the village, its square, you can see the comings and goings, the meeting of comadres and compadres at the end of the afternoon. It was in this *momentum* of participant observation that this affectivity stood out in relationships, in small conversations, in arranging meetings, in the messages and news that circulate through people[118] . These short- and long-range network relationships showed more than sociability or affection, they were in fact a *network of affections*. With this clue, we tried to map this network in order to find where it showed itself most clearly. We found the networks in the form of "help" (as they say) from the piem, as a counter-payment to the saint of devotion; in the communal mobilization of local and non-local actors (those from outside the village) who attend and effectively carry out the Ticumbi festivities, in a mixture of cooperation, religiosity and faith. In the work of the festeiros and festeiros obreiros, as the true *nodes of the network*. The network of affections is also presented through the specialists' practice of the art of healing, where these officials exercise their gift in return to the saint and to God, in the triple obligation: to give, to receive and to reciprocate. These actions of the gift create and recreate very strong bonds, moral bonds that cannot be undone. There is no payment, there is always a debt to the other.

Finally, in the sixth chapter, we present the ethnographic work itself, in order to understand how these healing officials, through their use of the knowledge of the art of healing, produce and reproduce the social space in the village. These specialists are recognized and legitimized as mediators between the saints and men, because they believe they have received a *gift* from God and exercise their office as a divine mission, in the sense of the Maussian gift. They are generally sought out to solve physical and spiritual problems, but they also offer advice. This knowledge does not compete with biomedical knowledge, but is articulated with it, and often warns and encourages devotees to seek medical help when they feel it is necessary. Diseases are thought of, analyzed and treated according to local gnosiology and therapies. Those illnesses that make sense within the local worldview are treated by the official of the art of healing, and those that don't make sense are referred to and treated by the health professionals at the medical center.

In local cosmology, there is no disjunction in understanding the world, because everything has an explanation, everything makes sense and if it doesn't, society tries to build one. Microcosm and macrocosm are thought of as united, where everything finds a correlation, and nothing happens above (heaven, world of the gods, saints) that doesn't happen below (earth, world of men), faith being what really heals. Although there are specialists, the knowledge to act in the world is laid out in a collective repertoire, which is transmitted diffusely through the generations. The relationship between society and nature is evident in this collective repertoire, and in the

118 This message network in Vila is very interesting and efficient. People send messages to each other.

body of knowledge of specialists and officials in the art of healing about the use of plants for medicinal purposes, the uses and secrets of plants for magical/religious purposes, added to the recognition of their world as a whole, as a single system.

In this way of thinking, we realize that specialists pass on their knowledge and arts within their own lineage[119] , and in general some child, grandchild or great-grandchild inherits and reproduces it. This is also the case with theTicumbi games. A congo master, a sailor or a secretary always takes his son to the game, and he ends up taking on the uniform within his father's group. This strategy of reproducing knowledge is very efficient in the village.

In fishing, knowledge is less direct, as it is learned from a friend, an uncle or a neighbor. The fisherman is not always the son of a fisherman, as the interlocutors told me. Artisanal fishing is practiced, but it is in serious danger of disappearing due to overfishing, which is depleting the fish supply. All the fishermen complained about the large trawlers that invade the village's communal maritime territory. These boats damage the fishermen's nets, scour the seabed, destroy the fishing grounds and leave a trail of destruction in their wake. Artisanal fishermen generally feel threatened and cornered, since the representatives of the fishing colony do little about the problem, IEMA takes no notice and the park manager has revealed that he has no power to deal with the matter.

Há serious problems of confrontation between the powers that be, between the state, as PEI and the Córrego Grande Biological Reserve (Rebio), the town hall and the local community. We would like to point out that it was not possible to describe the environmental issue and the conflict effectively, due to the analytical scope. However, the conflict is always muffled[120], withdrawn, silent, but it does exist. During the months in the field, there were many reports of arson in the Park, the Park Manager's disagreements with some members of the community, some arrests of teenagers from the village, an increase in drug trafficking and problems with the spread of drugs among teenagers in the village.

In short, the problems exist, and there is no efficient dialogue between the park and the local community. The level of understanding is very poor, and the policies undertaken by IEMA, through the PEI, do not meet the needs of the community. At most, it caters for the 'tourist' who visits the PEI, but little is designed for the town's community in general. Evidence of this is that most of the projects are carried out during the high season, when the village is full of tourists. We realize that we need to work on integrated projects and actions throughout the year.

There is also strong dissatisfaction with the health center and local health policies. This work doesn't touch on this issue clearly either, but the fact is clear from the interviews with some users of the service and health professionals. There is a big gap between biomedical knowledge and local knowledge.

The knowledge of midwives is not valued and is not being reproduced, although some (local) women have stated that they would like to learn, that it would be good if there were a course, and that it is very difficult to

---

119 There was even a concern to pass on this knowledge, as I heard from a benzedeira who was saddened because her son didn't want to continue his trade in the art of healing.

face the road to earn a baby in Conceiçao da Barra. The collective knowledge about plants for medicinal purposes is not recognized as such by the professionals at the health center or by the municipal health department. Uzelim (2005) has started an excellent project to disseminate knowledge, practices and use of plants for medicinal purposes[120] (appendix 1). The professionals at the health center (doctor, dentist and nurse) have had little involvement in this project, since the municipality, through the health department, has taken no interest in it. The PSF (Family Doctor Program) doesn't reach all the families in the village, and in informal conversations it becomes clear that the health agents not only don't receive the full amount of the federal government's budget, but are also hired rather than recruited, leaving them at the mercy of the mayors' political games.

Finally, it is necessary to close, between feelings, denunciations and propositions, bringing out the meaning of this intersubjective subject in the world. The world is constructed by him, invented, reinvented with its meanings and signifiers. The subject inhabits the world, which inhabits him. He geographizes the world, and in doing so inaugurates a place of his own. Through his actions in the world, in search of security, he delimits boundaries, proposes and exposes limits - he founds a territory. In doing so, he expresses an *ethos*, *language*[121] as a way of being in the world - *dasein* - being there, being. His actions in the world reveal his double dialectical face: individual and society. Being is metaphysical, complex, in a complex world, where it reveals its mythical, political, religious, psychological phase, in a given space/time.

All these final observations, the syntheses, will show us that the environmental issue in the Vila, with the confrontation between the community and the PEI, the issue of insufficient health policies, where the subject is neither known nor recognized, the funding issue of Angelim; in short, all these issues, as a whole, although they were not the focus of this work, are represented in that social space, and are interfering in this way of being in the local world. Many local actors fear 'progress', and others want 'progress'. Those who fear it do so because of the destruction of moral values that permeate relationships of friendship, family and cronyism, to the detriment of a negative capitalism that doesn't consider the place, installing its verticalities. And those who want it think about the benefits and convenience that paving the road would bring to life in the village, for example. That more jobs could be created and businesses could expand their profits.

So the question remains: what chance is this community being given to discuss these issues together, including with state political representatives? What is the future of Ticumbi as a religious association?

We need to draw up a social goal, a plan that makes it possible to include these people in the processes of building public policies. The issue, in our view, is first and foremost to understand and accept these local specificities as a *unique* way of being in the world, in order to make them subjects of conservation and not objects of it. There is an urgent need to open a channel of communication to foster a dialogic dialogue, a dialogue between knowledges.

May Geography, through its tools and methods, not only serve to wage war. May we use it to promote PEACE

120 Small talk: that is, what is said quietly, hidden away.
121 Language as in Gadamer (1985).

and dialogue.

8. Bibliographical references

ALVES, Isidoro Maria da Silva. *O carnaval devoto:* um estudo sobre a Festa de Nazaré em Belém. Petrópolis, RJ: Vozes, 1980.

ARAÙJO, Alceu Maynard. *Rustic Medicine.* Sao Paulo: Companhia Editora Nacional, 1961.

ARAÙJO, Melvina A. M.. *From medicinal herbs to phytotherapy* .Sao Paulo: Ateliê Editorial, 2002. 157p.

AYRES, José Ricardo de C. Mesquita. Subject, intersubjectivity and health practices. *Ciência e Saùde Coletiva*, Rio de Janeiro, V.6, n. 1, 2001. p. 63-72.

BAUMAN, Zygmunt. *Identity*. Rio de Janeiro: Jorge Zahar Ed., 2005. 108p.

. *Community:* the search for security in today's world. Rio de Janeiro:Jorge Zahar, 2003, p.7-39

BERGER, Peter L., LUCKMANN, Thomas. *The Social Construction of Reality*, 2ª ed.

Petrópolis: Vozes, 2002.

BERQUE, Augustin. *Écoumène*: introduction à L'étude dès milieux humains. Paris: BELIN, 2000.

. Landscape-mark, landscape-matrix: problematic elements for a cultural geography. In: CÔRREA, Roberto L.; ROSENDAHL, Zeny (eds). *Paisagem, tempo e cultura.* 2° ed. Rio de Janeiro: EdUERJ, 2004. p. 84-91.

BOLTANSKI, L. *Social classes and the body*. Rio de Janeiro: Graal, 1989. 185p.

BONNEMAISON, Joël. Journey around the territory. In: CÔRREA, Roberto L.; ROSENDAHL, Zeny (eds). *Cultural geography:* a century (3). Rio de Janeiro: EdUERJ, 2002, p.83-131

BOURDIEU, Pierre. *The economy of symbolic exchanges.* Sao Paulo: Editora Perspectiva, 1987.

. *The symbolic power*. Rio de Janeiro: Bertrand, 1989.

. *Practical Reasons:* on the theory of action. Sao Paulo: Papirus, 1997

BRANDÂO, Carlos Rodrigues. *Plantar, colher, comer:* um estudo sobre campesinato goiano. Rio de Janeiro: Ediçôes Graal, 1981. 181p.

BRAZIL, Constitution of. *Federative Republic of Brazil*. Brasilia: Senado Federal, Centro Gràfico, 1988. 262p.

BUTTIMER, Anne. Apprehending the dynamism of the lived world. In: CRISTOFOLETTI, Antonio (org). *Perspectives on Geography.* Sao Paulo: DIFEL, 1982.

BOURDIEU, Pierre. O poder *simbòlico.* 2° ed. Rio de Janeiro: Bertrand Brasil, 1998.

. *Practical Reasons*. Campinas, SP: Papirus, 1996.

BRANDÂO, Carlos Rodrigues. *The feast of the saint in black*. Rio de Janeiro; FUNARTE/Instituto Nacional do Folclore; Goiânia: Universidade Federal de Goiàs, 1985.

CAILLÉ, Alain. *Anthropology of the Gift*: the third paradigm. Petrópolis, RJ: Vozes, 2002.

The gift of words: what saying intends to give. In:MARTINS, Paulo Henrique (org). *The gift among the moderns:* discussion on the foundations and rules of the social. Petrópolis: 2002. p.99-136.

CAIRUS, Henrique. The corpus hippocraticum. IN: CAIRUS, H.; RIBEIRO JR., Wilson A.. *Hippocratic texts:* the sick, the doctor and the disease. Rio de Janeiro: Editora Fiocruz, 2005, p. 25-38.

CALABRIA, Juliana; SILVA, José Carlos Gomes da. *Social network and ritual process:* the double face of Congada in Uberlândia - MG. Available atwww.propp.ufu.br/revistaeletronica/edicao2004/humanas/rede_social.PDF

Accessed on: April 20, 2008.

CARDOSO DE OLIVEIRA, Roberto. *O trabalho do antropòlogo*. 2° ed. Brasilia: Paralelo 15, Sao Paulo: Ed. UNESP, 2000.

CARNEIRO, Sandra de Sà. Conference presented at: Trails of the Sacred: the Hieropolis. *VI National Symposium and II International Symposium on Space and Culture, October 29-31, 2008.* Rio de Janeiro: NUPEC-UERJ.

CARVALHO, Màrcia Siqueira de. Terra (In) cognitae. In: SEEMANN, Jorn (org). *A aventura cartogràfica:* perspectivas e reflexões sobre a cartografia humana. Fortaleza: Expressao Gràfica e Editora, 2006. p.75-86.

CASTELLS, Manuel. *The power of identity*. Vol. II. 3ª ed. Sao Paulo: Ed. Paz e Terra, 2001, p.18-30; 142-168;417-427.

CASTORIADIS, Cornelius. The institution of society and religion. In: _________.(org) *The destinies of totalitarianism and other writings.* Porto Alegre: L&PM, 1985.

CERTAU, Michel de; GIARD, Luce; MAYOL, Pierre. *The invention of everyday life*. 2- living, cooking. 5th ed. Petrópolis, RJ: Vozes, 1996. p. 37-69; 203-207.

. *The invention of everyday life.* 1- Arts of making. Petrópolis, RJ: Vozes, 2000.

CLAVAL, Paul. *Territory and post-modernity*. Geo-graphy Magazine - Year 1, No. 2, 1999.

. The French contribution to the development of the cultural approach in geography. In: CORREA, Roberto L.; ROSENDAHL, Zeny (eds). *Introduction to cultural geography*. Rio de Janeiro: Bertrand Brasil, 2003. p.147-166.

. The landscape of geographers. In: CORRÊA, Roberto L.; ROSENDAHL, Zeny (eds). *Landscapes, texts and identity*. Rio de Janeiro: EdUERJ, 2004. p.13-74

CLAVAL, Paul. *Cultural geography*. 3ª ed. Florianópolis: Ed. da UFSC, 2007.

CORRÊA, Roberto Lobato. Space: a key concept in Geography. In: CASTRO, Elias de; GOMES, Paulo César da C.; CORRÊA, Roberto Lobato. *Geography:* concepts and themes. Rio de Janeiro: Bertrand Brasil, 1995, p.15-47.

CORRÊA, Aureanice de Mello. Festa da Irmandade da Boa Morte: the dispute over its meaning. In:

ROSENDAHL, Zeny; CORRÊA, Roberto Lobato (eds). *Space and culture:* thematic plurality. Rio de Janeiro: EdUERJ, 2008. p. 249 - 278.

COSGROVE, Denis E. Towards a radical cultural geography: problems of theory. In: CORREA, Roberto L.; ROSENDAHL, Zeny (eds). *Introduction to cultural geography*. Rio de Janeiro: Bertrand Brasil, 2003. p. 103-134.

COUTO, Patricia Brandao. Feast of the Rosary. *Iconography and poetics of a rite*. Niterói: EDUFF, 2003.

DA MATTA, Roberto. *Carnival as a rite of passage*. In: Ensaios de antropologia estrutural. Rio de Janeiro: Vozes, 1973. p.121-168.

. *Carnival, scoundrels and heroes*. 6th ed. Rio de Janeiro: Rocco, 1997.

DIAS, Leila Christina. The meanings of the network: notes for a discussion. In: DIAS, L. C. & SILVEIRA, Rogério Leandro da. (orgs), *Redes, sociedades, territórios*. 2ed. Santa Cruz do Sul: EDUNISC, 2007. p.11-28.

DI MÉO, Guy. *La géographie en fêtes*. Paris : OPHRYS, 2001.

DUNKAN, James S..The supra-organic in American cultural geography. In: CORRÊA, Roberto L.; ROSENDAHL, Zeny (eds). *Introduction to cultural geography*. Rio de Janeiro: Bertrand Brasil, 2003. p. 63-102.

DURANT, Will. *The history of philosophy.* Sao Paulo: Ed. Nova Cultural, 1996. Chap. VII, p. 285-327.

ELIADE, Mircea. *Images and Symbols: an* essay on mystical-religious symbolism. Sao Paulo: Martins Fontes, 1991.

. *The sacred and the opprobrious:* the essence of religions. Sao Paulo: Martins Fontes, 1992.

ELIAS, Norbert. *The society of individuals*. Rio de Janeiro: Jorge Zahar Ed., 1994. 201p.

ELIAS, Norbert e SCOTSON, John L. *Os estabelecidos e os outisiders*: sociologia das relações de poder a partir de uma pequena comunidade. Rio de Janeiro: Zahar Editor, 2000.

FERREIRA, Aurélio Buarque de Holanda. *Novo Aurélio XXI:* dicionàrio de lingua portuguesa. 3rd ed. Rio de Janeiro: Nova Fronteira, 1999.

FERREIRA, Simone R. Batista. *From plenty to scarcity*: the cellulose agro-industry and the end of communal territories in the Far North of Espirito Santo. Master's dissertation in Human Geography. Sao Paulo: USP, 2002.

FOCAULT, Michel. The subject and power. In: RABINOW, Paul; DREYFUS, Hubert L. *Michel Foucault, uma trajectória filosófica* (para além do estruturalismo e da hermenêutica). Rio de Janeiro: Forense Universitària, 1995. p.231-249.

. *Summary of courses at the Collège de France*. Rio de Janeiro: Jorge Zahar Ed., 1997. p. 9-24.

. *The archeology of knowledge*. 7th ed. Rio de Janeiro: Forense Universitària, 2004.

. *Microfisica do Poder*. 23ª ed. Rio de Janeiro: Ediçôes Graal, 2007.

FREIRE, Paulo. *Pedagogy of the oppressed.* 1st ed. Rio de Janeiro: Paz e Terra, 1987. 184p.

FREYRE, Gilberto. *Casa-grande e senzala:* formaçao da familia brasileira sob o regime de economia patriarcal. Rio de Janeiro: Livraria José Olympo Editora, 1954.

GADAMER, Hans-Georg. *The actuality of beauty:* art as a game, a symbol of celebration. Rio de Janeiro: Tempo brasileiro, 1985.

GARCIA, Sylvia Gemignani. Folklore and sociology in Florestan Fernandes. *Tempo Social*; Rev. Sociol. USP, S. Paulo, 13(2): 143-167, November 2001.

GEERTZ, Clifford. *The interpretation of cultures.* Rio de Janeiro: LTC, 1989. 323p.

. *Local knowledge.* Petrópolis, RJ: Vozes, 2002. 366p.

. *New light on anthropology.* Rio de Janeiro: Jorge Zahar Editores, 2001. 247p.

GABARRA, Larissa. *Congado*: religion and power in Minas Gerais in the 19th century. Full paper - History Symposium Proceedings. Santa Catarina, 2007.

GODBOUT, Jacques T. *O espirito da dàdiva.* Rio de Janeiro; Editora da Fundaçao Getúlio Vargas, 1999.

GODELIER, Maurice. *The enigma of the gift.* Rio de Janeiro: Civilizaçao Brasileira, 2001.

GUEDES, Simone Lahud. Cases of divine healing and the construction of difference. In: Anthropological Horizons. *Body, illness and health.* UFRGS/IFCH. Postgraduate Program in Social Anthropology. Year 1, n.1 (1995). Porto Alegre: PPGAS, 1998.

GUIDDENS, Anthony. *The consequences of modernity.* Sao Paulo: Editora UNESP, 1991.

HAESBAERT, Rogério. *O mito da desterritorialização*: do fim dos territórios à multiterritorialidade. Rio de Janeiro: Bertrand Brasil, 2004.

HALL, Stuart. *Cultural identity in postmodernity.* 4th ed. Rio de Janeiro: Ed. DP&A, 2000. 102p.

. Who needs identity? In: Silva, Tomaz Tadeu da.(org.) *Identidade e diferença:*a perspectiva dos estudos culturais. Petrópolis, RJ: Ed. Vozes, 2000 a. p.103-133.

HARAWAY, Donna. Cyborg manifesto: science, technology and feminist-socialism at the end of the 20th century. In: SILVA, Tomaz Tadeu da (org). *Anthropologia do ciborque* - as vertigens do pós-humano. Belo Horizonte: Autêntica, 2000. p. 37-129

HARVEY, David. *The postmodern condition.* 14ª ed. Sao Paulo: Ed.Loyola, 2005. p.185-289.

HEIDEGGER, Martin. *Being and Time I.* Petrópolis, RJ: Vozes, 1995.

. *Essays and conferences.* Petrópolis, RJ: Vozes, 2002.

HILMANN, James. *City & Soul* - Sao Paulo: Stùdio Nobel, 1993.174 p.

HOBSBAWN, Eric; RANGER, Terence. *The invention of traditions.* Sao Paulo: Paz e Terra, 2002.

HOLZER, Werther. Humanist *geography - its trajectory from 1950 -1990.* Master's dissertation. Rio de

Janeiro: UFRJ/ PPGG, 1992.

. *Humanist geography*: a review. In: *Espaço e Cultura*. n° 3, Rio de Janeiro: UERJ, 1996, p. 8-19.

. *A phenomenological study of landscape and place:* the chronicle of travelers in 16th century Brazil. PhD thesis. Sao Paulo: USP. 1998.

. Eric Dardel's phenomenological geography. In: ROSENDAHL, Zeny; CÔRREA, Roberto Lobato. *Matrices of cultural geography*. Rio de Janeiro: EdUERJ, 2001, p.103-122

. The costume: theoretical reflections on the vernacular landscape. In: ROSENDAHL, Zeny; CORRÊA, Roberto Lobato (eds). *Space and culture:* thematic plurality. Rio de Janeiro: EdUERJ, 2008. p. 155-172.

HOLZER, Werher; HOLZER, Selma. Cartography for children: where do they belong? In:

SEEMANN, Jorn (org). *The cartographic adventure:* perspectives and reflections on human cartography. Fortaleza: Expressao Gràfica e Editora, 2006. p.201-217.

JOGAIB, Alexandre de Oliveira. *The territorial (re)production of Espirito Santo:* from the 16th to the 21st century. Monograph presented to the Geography Department, Geosciences Institute, Fluminense Federal University, Niterói, 2005. 81p.

KOLZER, Salete. Communicating and representing: maps as socio-cultural constructions. In: SEEMANN, Jorn (org). *The cartographic adventure:* perspectives and reflections on human cartography. Fortaleza: Expressao Gràfica e Editora, 2006. p.131-149.

LARAIA, Roque de Barros. *Culture:* an anthropological concept. 16th ed. Rio de Janeiro: Ed. Zahar, 2003. 117p.

LÉVI-STRAUSS, Claude. *The sorcerer and his magic.* Structural anthropology. Rio de Janeiro, 1970. p. 183-203.

. *Symbolic efficiency*. Structural anthropology. Rio de Janeiro: Tempo Brasileiro, 1975. p. 215-236.

. *O pensamento selvagem*. 2ª ed. Campinas, SP: Papirus, 1997.

LIMA, Deborah (coord.), OLIVEIRA, Fernanda C. de, MARQUES, Carlos E., FARIA, Ana Tereza, BARBI, Rafael. *Anthropological Report on the Historical, Economic and Socio-Cultural Characterization of Quilombo deMumbuca, Lower Jequitinhonha, Minas Gerais*. Center for the Study of Quilombola and Traditional Populations. NUQ . Federal University of Minas Gerais, August 2007. CD-ROM.

LYNCH, Kevin. *The image of the city*. Sao Paulo: Martins Fontes, 1999.

LOPES DE SOUZA, Marcelo José. Territory: on space and power, autonomy and development. In: CASTRO, Elias de; GOMES, Paulo C. da Costa: CÔRREA, Roberto L.(orgs). *Geography*: concepts and themes. Rio de Janeiro: Bertrand Brasil, 1995. p.77-115.

LOTT, Henrique Marques. *Art and religion in Gadamer's hermeneutics*. Master's dissertation. PG Science of Religion. Juiz de Fora: UFJF, 2007.

MALUF, Ued. *Culture and Mosaic: an* introduction to the theory of strangeness. 2nd ed. 169 p.

MARQUARD, Odo. Una pequena filosofia de la fiesta. In: SCHULTZ, Uwe. *La fiesta.* Barcelona/ Madrid: Altaya, 1998, p.356-367.

MARTINS, Paulo Henrique. *The gift among the moderns:* discussion on the foundations and rules of the social. Petrópolis, RJ: Vozes, 2002.

MAUSS, Marcel. *Sociology and anthropology*. Vol. II. Sao Paulo: Edusp, 1974.

MEDEIROS, Rogério. The women's rebel Ticumbi. *Seculàrio diàrio.* March 12-13, 2005 edition, Vitória- ES. Available at:< http://www.seculodiario.com.br/arquivo/2005/marco/12 13/reportagens/reportagens/12 03 01.asp> Accessed on September 7, 2008.

MELO, Vera Lùcia M. de Oliveira. *Landscape from the perspective of new geographical approaches.* Proceedings of the X Meeting of Latin American Geographers - March 20-26, 2005, Sao Paulo, USP.

MORAES, Antonio Carlos Robert. *Geografia* : pequena história critica. 20ª ed. Sao Paulo: Annablume, 2005.

MOREIRA, Roberto José. Ruralities and globalization: rehearsing an interpretation. In: .(org) CARNEIRO, Maria José. *et.al.*. *Social identities*: ruralities in contemporary Brazil. Rio de Janeiro: DP&A, 2005. p. 15-40.

MOREIRA, Ruy. *Thinking and being in geography*: essays on the history, epistemology and ontology of geographical space. Sao Paulo: Contexto, 2007.

. Spaciousness: a reflection on the problem of the ontology of space. In: OLIVEIRA, M. Pinon, et al. *O Brasil, a América Latina e o mundo:* espacialidades contemporâneas (I). Rio de Janeiro: Lamparina, 2008.

MORIN, Edgar. *Method 3:* the knowledge of knowledge. Porto Alegre: Sulina, 1999.

MOSCOVICI, Serge. *The god-making machine*. Rio de Janeiro: Imago, 1990.

MOTTA-MAUÉS. Maria Angélica. *Workers and comrades:* gender relations, symbolism and ritualization in an Amazonian community. Belém: Editora Universitària UFPA, 1993. 216p.

MOTA, Christine V. R.; FERREIRA, Solange L.; CORRÊA, Aurikson. *Winds that bring destruction and beauty.* Sao Mateus, ES: Opçâo, 1998.

MUSSO, Pierre. Le réseau: de La mythologie grecque à l'idéologie d'Internet. In: MUSSO (org). *Networks and society*. Paris: Univesitaires de France (PUF), 2003.p.15-42.

NEVES, Delma Pessanha. *Miraculous cures and the idealization of social order*. Niterói, Universidade Federal Fluminense - CEUFF- PROED, 1984 .64p.

OLIVEIRA, Sonia Acioli. The construction of knowledge and practices of popular groups in the field of Collective Health. In: *Tecendo saberes* - Jornada de Pesquisadores em Ciências Humanas, UFRJ. Organized by CFCH, UFRJ, with the support of the José Bonifacio University Foundation. Available in CR-ROOM. Rio de Janeiro, December 6th 2000.

PEI- Itaûnas State Park. *Institutional Diagnosis of PEI*: Integration Plan with the surroundings of Itaúnas State

Park, ES. December 2006.

PEIRANO, Mariza. *Rituals yesterday and today.* Rio de JaneiroJorge Zahar Ed. , 2003.

QUINTANA, Alberto M. *A ciência da benzedura:* mau olhado, simpatias e uma pitada de psicanalise. Bauru, SP: EDUSC, 1999.

PLATO. *Timaeus - Critias - The second Alcaebiades - Hypias Minor.* Translated by Carlos Alberto Nunes. 3ª rev. ed. Belém: EDUFPA, 2001.

PORTO-GONÇALVES, Carlos Walter. *From Geography to Geo-graphies: a world in search of new territorialities.* Revista Clasco, II Conferencia Latinoamericana y Caribena de Ciencias Sociales. University of Guadalajara, Mexico, November 21-22, 2001.

. *The geography of the social.* Paper presented at the International Seminar "Conflicto Social, Militarizacion y Democracia en América Latina - nuevos problemas y desafios para los estrudios sobre conflicto y paz en la region", Consejo Latino Americano de iencias Sociales - Clacso - and Agencia Sueca de Desarrollo Internacional - Asdi. Buenos Aires, Argentina, September 16-18, 2002.

QUEIROZ, Maria Isaura P. de. *Bairros rurais paulistas: dinàmica das relações bairro rural- cidade.* Sao Paulo: Livraria Duas Cidades, 1973, pp. 3-9; 12-145.

. Dialectics of rural and urban: Brazilian examples. In: BLAY, Eva Alterman (org.). *A luta pelo espaço.* 2a ed. Petrópolis: Editora Vozes, 1979. p.23-73.

RABINOW, Paul; DREYFUS, Hubert L. Poder e verdade..*Michel Foucault, uma trajetoriaafilosòfica* (para além do estruturalismo e da hermenêutica). Rio de Janeiro: Forense Universitaria, 1995.p. 202-224.

RAFFESTIN, Claude. *For a geography of power.* Sao Paulo: Editora Atica S.A., 1993.

RODRIGUES, *Antonio Greco. Searching for roots. In: Anthropological Horizons.* Nature and culture. UFRGS/IFCH. Postgraduate Program in Social Anthropology. Year 7. n.16 (2001). Porto Alegre: PPGAS, 2001.p.131-144. p.47-62.

ROSENDAHL, Zeny. The paths of theoretical construction: ratifying and exemplifying the relationship between space and religion. In: CÔRREA, Roberto L.; ROSENDAHL, Zeny (eds). *Space and culture*: thematic plurality. Rio de Janeiro: EdUERJ, 2008 a. p.47-77.

. The dimension of the sacred place: ratifying the dominance of emotion and the feeling of being-in-the-world. In: OLIVEIRA, Màrcio Pinon; COELHO, Maria Célia; CORREA, Aureanice de Mello (eds). *Brazil, Latin America and the world:* contemporary spatialities (II). Rio de Janeiro:Lamparina, Faperj, Anpege, 2008 b. p.331-337.

SACK, R. D.. *Human Territoriality.* Cambridge: Cambridge University Press, 1986.

SANTOS, Boaventura Sousa. Towards a sociology of absences and a sociology of emergencies. Available at: http://www.ces.fe.uc.pt/bss/index.php. Accessed on December 1, 2003. [Published in *Revista Critica de Ciências Sociais*, 63 - October 2002]

SANTOS, Maria da Graça M. Poças. Sanctuaries as places of construction of the sacred and of hierophanic memory: outline of a typology. In: CORRÊA, Roberto L.;

ROSENDAHL, Zeny (eds). *Space and culture: thematic plurality.* Rio de Janeiro: EdUERJ, 2008, p.79-104.

SANTOS, Milton. *Por uma geografia nova* : da critica da geografia a uma geografia critica. 6th ed. Sao Paulo: Editora da USP, 2004.

. *The nature of space - technique and time/reason and emotion.* Sao Paulo: Hucitec, 2004 a. 384p.

SAUER, CARL O.. A educação de um geógrafo.In: *Geographia,* Revista do Programa de PG em Geografia da UFF.Ano II, n° 4, Niterói: UFF/EGG, 2001. p. 135-150.

. Landscape morphology. In: Côrrea, Roberto Lobato, Rosendahl. *Paisagem, tempo e cultura.* 2° ed. Rio de Janeiro: edUERJ, 2004. p. 12-73.

SCHERER-WARREN, I. Redes Sociales y de movimientos en la sociedad de la información. *Revista Nueva Sociedad,* Caracas, n. 196, p. 77-92, Mar-Apr, 2005.

. The politics of social movements for the rural world. *II Meeting of the Rural Studies Network,* Panel III: rural issues and politics, held at the IFCS/UFRJ, Rio de Janeiro, 11-14/09/2007.

Social networks: trajectories and frontiers. In: DIAS, Leila Christina; SILVEIRA, Rogério Leandro Lima da. (eds). *Networks, societies and territories.* Santa Cruz do Sul, RS: UNISC, 2007.

SCHUCK, Rogério José - *Através da compreensão da historicidade para uma historicidade da compreensão como apropriação da tradiçto.* PUC RGSul - PhD thesis - Faculty of Philosophy of the Faculty of Philosophy and Human Sciences of the Pontifical Catholic University of Rio Grande do Sul.Porto Alegre:PUC, 2007.

SEEMANN, Jom. Imaginary lines in cartography: the invention of the first meridian. In: (org). *The cartographic adventure:* perspectives and reflections on human cartography.

Fortaleza: Expressao Gràfica e Editora, 2006. p.111-129.

. Approach to Brazilian Cultural Geography. Panel: history, theory and methods in cultural geography. *Presented at the VI National Symposium and II International Symposium on Space and Culture, October 29-31, 2008.* Rio de Janeiro: NEPEC- UERJ, 2008.

SIGAUD, Lygia. The vicissitudes of the "Essay on the Gift". *Mana,* vol. 5. Rio de Janeiro, October 1999. p.1-27. Available at:www.scielo.br/pdf/mana/v5n2/v5n2a04.pdf . Accessed on: September 21, 2005.

SILVA, Jailson de Souza e. *The plurality of identities in the Maré neighborhood of Rio de Janeiro.* Reflection text for the course Identity, territory and public policies. 2nd semester 2005. Postgraduate course in Geography. Institute of Geosciences, Fluminense Federal University. Niterói, RJ.

SILVA, Joseli Maria. Culture and urban territorialities: a small town approach.

Revista de História Regional, Vol. 5- n° 2- 2000. Available atwww.uepg.br/rhr/v5n2/joseli.htm , accessed on 30/03/2006.

SILVA, Tomâs Tadeu da. *Identity and Difference*: the perspective of cultural studies. Silva, Tomâs Tadeu da (org). Hall, Stuart; Woodward, Kathryn. Petrópolis, RJ; Vozes, 2000.

. We, cyborgs: the electric body and the dissolution of the human. SILVA, Tomaz T. da (org); Kunzru, Hari; Haraway, Donna. IN: *Anthropologia do ciborgue: as vertigens do pós- humano*.Belo Horizonte: Autêntica, 2000 a. p. 11-17

SODRÉ, Muniz. *O terreiro e* a *cidade:* a forma social negro-brasileira. Petrópolis: Vozes, 1983. 165p.

STEIL, Carlos Alberto. Catholicism and culture. In: VALLA, Victor Vincent (org). *Religion and popular culture.* Rio de Janeiro: DP&A, 2001.p.9- 40.

TUAN, Y-Fu. *Topophilia.* Sao Paulo: DIFEL, 1980.

. *Space and place*: the perspective of experience. Sao Paulo: DIFEL, 1983.

TURNER, Victor. *The ritual process.* Petrópolis: Editora Vozes, 1974.

WEBER, Max. *Economy and Society*: Foundations for a Comprehensive Sociology. Vol. I. Brasilia, DF: Ed. UNB: Sao Paulo: Imprensa Oficial do Estado de SP, 2004, 464p.

WOODWARD, Kathryn. Identity and difference; a theoretical and conceptual introduction. In: Silva, Tomaz T. da; Hall, Stuart; Woodward, K.(eds) *Identity and difference: a cultural studies perspective.* 4ª ed. Petrópolis: Ed. Vozes, 2005, p. 7-72.

UZELIN, Patricia. *The use of home remedies in the village of Itaûnas.* Monograph presented to the Department of Agriculture of the Federal University of Lavras, specialist course in Medicinal Plants: Management, Use and Manipulation. Lavras, 2005.

VALLA, Victor Vincent. What does health have to do with religion? In: VALLA, Victor Vincent (org). *Religion and popular culture.* Rio de Janeiro: DP&A, 2001.p.113-139.

XAVIER, Maria Aparecida de Sà. *Study of the Symbolic Representations of Health/Disease/Healing in the community of Saco do Mamanguà, Paraty, RJ.* Master's dissertation, PGCA, UFF, 2004.181p.

. *Health, disease and medicine in a caiçara fishing community in Brazil.* Collection

FazereSaberes, n° 2, directed by Ana Luisa Janeira. Lisbon, PT: Apenas Livros Ltda, 2005, 67 p.

. *Traditional knowledge as an expression of territoriality:* the case of the caiçara community of Saco do Mamanguà, Paraty, RJ; 2006 (unpublished)

ZALUAR, Alba. *The men of God:* a study of saints and festivals in popular Catholicism. Rio de Janeiro: Zahar Editores, 1983.

9. Annexes

9.1 Home Remedies from Itaûnas: Cia Oficios da Terra.

APRESENTAÇÃO

Olá. Nas páginas a seguir, você vai encontrar um pouco de nós, moradoras e moradores da Vila de Itaúnas- lugarejo de natureza atraente no extremo norte capixaba. Aqui vai o resultado de vários meses de trabalho no ano de 1998 com remédios caseiros à base de ervas medicinais.

Foram cinco encontros onde, através da troca de experiências, produzimos vários remédios, discutimos fórmulas, conhecemos um pouco da flora nativa e exótica; comemos bolos, tomamos chás e rimos um bocado da nossa própria sorte.

O grupo - que já foi semente, brotou e tornou-se muda disposta a crescer - segue seu caminho na busca de qualidade de vida para cada um e para todos que nos cercam, sejam filhos, amigos, amores ou simplesmente espíritos convergentes com a idéia de que viver bem é possível, praticável e muito gostoso.

Cada informação desta cartilha é fruto destes encontros saudáveis e felizes, de um grupo aberto (braços e coração) para crescer.

Bem vindo à

Cia. Ofícios da Terra

ESCOLA LIVRE PARA OFÍCIOS DA TERRA

entendida como corrente filosófica, isto é, como um ideal que nos guia nas ações. Ela é livre para poder ser tão criativa quanto seus alunos e mestres, que tem garantida a liberdade de mudar de papéis, nela o mestre vira aluno para o aluno ser mestre da experiência que carrega. Nossa escola é da **Cidadania e Meio Ambiente**, pois os alunos e professores se aprimoram na arte de ser cidadãos capazes de um meio ambiente saudável para todos os seres: animais, vegetais ou minerais, nativos de todos os "cantos" da Terra.

E assim nasce a *Escola Livre de Cidadania e Meio Ambiente*, formando cidadãos para todos os Ofícios da Terra e nosso trabalho com o resgate do uso de ervas medicinais dentro dessa Escola, nada mais é do que uma semente germinando em terra fértil, em busca de abundância para todos.

INFORMATIVO

Esta cartilha é apenas a publicação dos dados obtidos nos encontros sobre o uso de remédios caseiros, realizados no decorrer de 98, na comunidade de Itaúnas, tendo como objetivo a preservação do conhecimento empírico da população , sobre o tema, o resgate e a transmissão dessa cultura. O presente trabalho não tem cunho científico, pois se trata do registro de depoimentos e vivências do grupo envolvido nesta atividade.

ERRATA

Algumas falhas passaram despercebidas pela revisão. *A Cia. Ofícios da Terra conta com a sua compreensão.*

Página 6 - na fórmula Pomada, o óleo não é aquecido, a mistura óleo e tintura acontece à temperatura ambiente.

Página 8 - onde se lê dessecadas leia-se secas.

Página13- na receita "Pó para anemia" , as folhas de mandioca não devem ser levadas ao forno.

Página15- o título "Garrafada da Dona Doróta" deve ser substituído por "Óleo de Côco".

Página16- onde se lê Mdicnais leia se Medicinais. O Alecrim citado como calmante em banhos refere-se ao Alecrim de Caboclo, nossa Alfazema nativa.

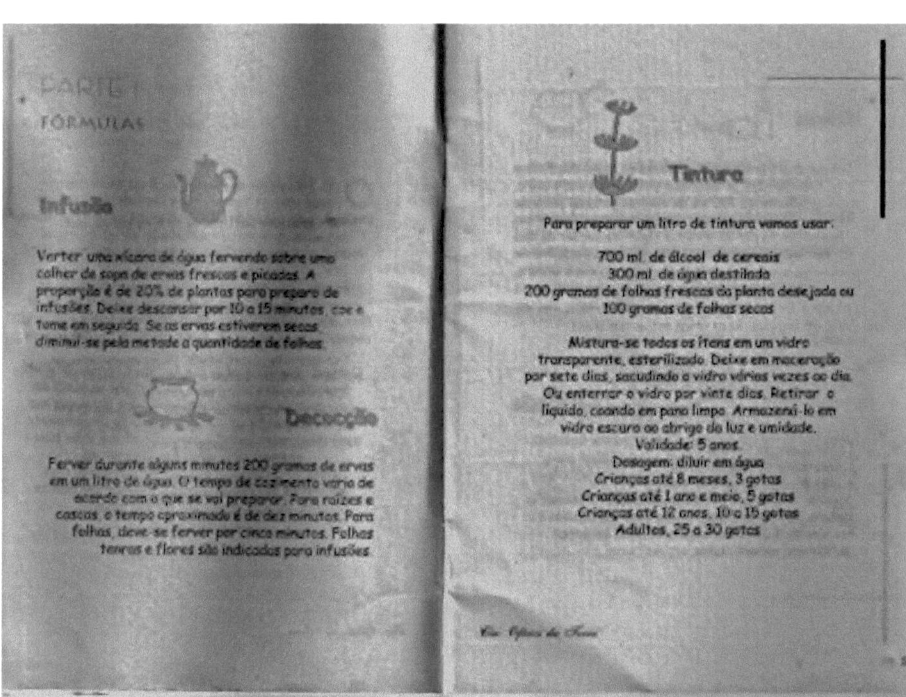

PARTE I
FÓRMULAS

Infusão

Verter uma xícara de água fervendo sobre uma colher de sopa de ervas frescas e picadas. A proporção é de 20% de plantas para preparo de infusões. Deixe descansar por 10 a 15 minutos, coe e tome em seguida. Se as ervas estiverem secas, diminui-se pela metade a quantidade de folhas.

Decocção

Ferver durante alguns minutos 200 gramas de ervas em um litro de água. O tempo de cozimento varia de acordo com o que se vai preparar. Para raízes e cascas, o tempo aproximado é de dez minutos. Para folhas, deve-se ferver por cinco minutos. Folhas tenras e flores são indicadas para infusões.

Tintura

Para preparar um litro de tintura vamos usar:

700 ml. de álcool de cereais
300 ml. de água destilada
200 gramas de folhas frescas da planta desejada ou
100 gramas de folhas secas

Mistura-se todos os itens em um vidro transparente, esterilizado. Deixe em maceração por sete dias, sacudindo o vidro várias vezes ao dia. Ou enterrar o vidro por vinte dias. Retirar o líquido, coando em pano limpo. Armazená-lo em vidro escuro ao abrigo da luz e umidade.

Validade: 5 anos.
Dosagem: diluir em água
Crianças até 8 meses, 3 gotas
Crianças até 1 ano e meio, 5 gotas
Crianças até 12 anos, 10 a 15 gotas
Adultos, 25 a 30 gotas

5

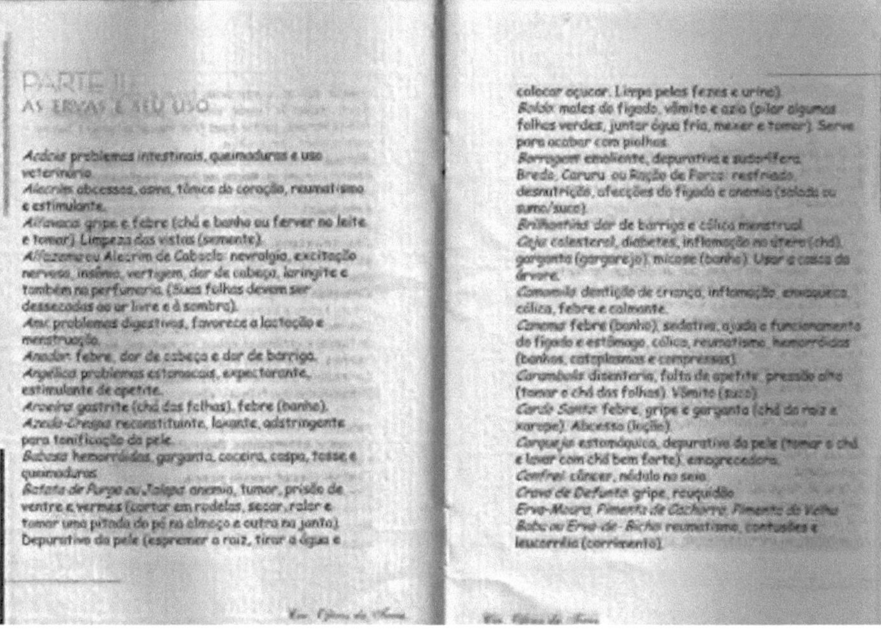

PARTE II
AS ERVAS E SEU USO

Acácia: problemas intestinais, queimaduras e uso veterinário.
Alecrim: abcessos, asma, tônico do coração, reumatismo e estimulante.
Alfavaca: gripe e febre (chá e banho ou ferver no leite e tomar). Limpeza das vistas (semente).
Alfazema ou Alecrim de Caboclo: nevralgia, excitação nervosa, insônia, vertigem, dor de cabeça, laringite e também na perfumaria. (Suas folhas devem ser dessecadas ao ar livre e à sombra).
Anis: problemas digestivos, favorece a lactação e menstruação.
Anador: febre, dor de cabeça e dor de barriga.
Angélica: problemas estomacais, expectorante, estimulante de apetite.
Aroeira: gastrite (chá das folhas), febre (banho).
Azedo-Chaga: reconstituinte, laxante, adstringente para tonificação da pele.
Babosa: hemorróidas, garganta, coceira, caspa, tosse e queimaduras.
Batata de Purga ou Jalapa: anemia, tumor, prisão de ventre e vermes (cortar em rodelas, secar, ralar e tomar uma pitada do pó no almoço e outra na janta). Depurativo da pele (espremer a raiz, tirar a água e colocar açúcar. Limpa pelas fezes e urina).
Boldo: males do fígado, vômito e azia (pilar algumas folhas verdes, juntar água fria, mexer e tomar). Serve para acabar com piolhos.
Borragem: emoliente, depurativa e sudorífera.
Bredo, Caruru ou Ração de Porco: resfriado, desnutrição, afecções do fígado e anemia (salada ou sumo/suco).
Brilhantina: dor de barriga e cólica menstrual.
Caju: colesterol, diabetes, inflamação no útero (chá), garganta (gargarejo), micose (banho). Usar a casca da árvore.
Camomila: dentição de criança, inflamação, enxaqueca, cólica, febre e calmante.
Canema: febre (banho), sedativo, ajuda o funcionamento do fígado e estômago, cólica, reumatismo, hemorróidas (banhos, cataplasmas e compressas).
Carambola: disenteria, falta de apetite, pressão alta (tomar o chá das folhas). Vômito (suco).
Cardo Santo: febre, gripe e garganta (chá da raiz e xarope). Abcesso (loção).
Carqueja: estomáquica, depurativa da pele (tomar o chá e lavar com chá bem forte), emagrecedora.
Confrei: câncer, nódulo no seio.
Cravo de Defunto: gripe, rouquidão.
Erva-Moura, Pimenta de Cachorro, Pimenta de Velho, Bobo ou Erva-de-Bicho: reumatismo, contusões e leucorréia (corrimento).

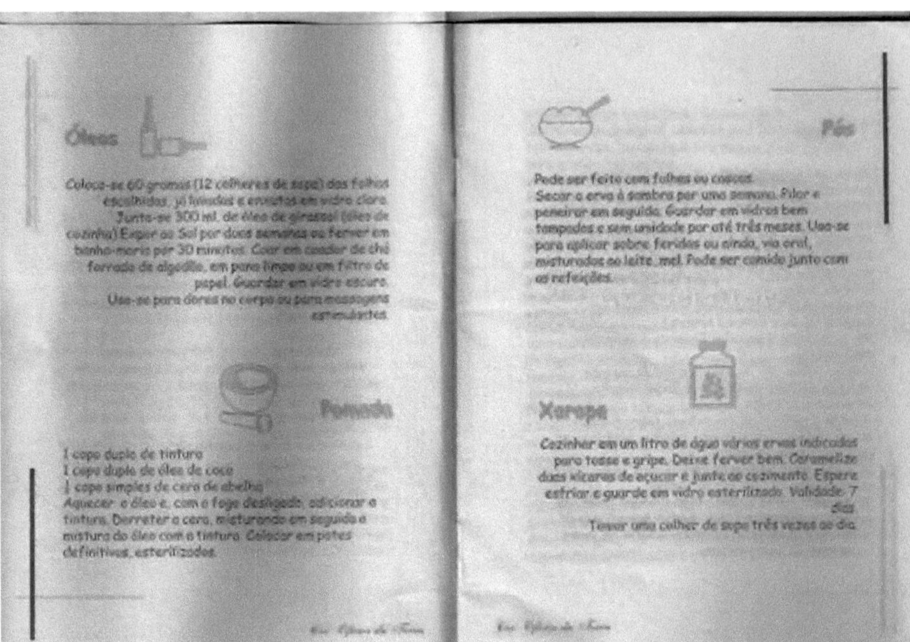

Óleos

Coloca-se 60 gramas (12 colheres de sopa) das folhas escolhidas, já lavadas e enxutas em vidro claro. Junta-se 300 ml. de óleo de girassol (óleo de cozinha). Expor ao Sol por duas semanas ou ferver em banho-maria por 30 minutos. Coar em coador de chá forrado de algodão, em pano limpo ou em filtro de papel. Guardar em vidro escuro.
Usa-se para dores no corpo ou para massagens estimulantes.

Pomada

1 copo duplo de tintura
1 copo duplo de óleo de coco
½ copo simples de cera de abelha
Aquecer o óleo e, com o fogo desligado, adicionar a tintura. Derreter a cera, misturando em seguida a mistura do óleo com a tintura. Colocar em potes definitivos, esterilizados.

Pós

Pode ser feito com folhas ou cascas.
Secar a erva à sombra por uma semana. Pilar e peneirar em seguida. Guardar em vidros bem tampados e sem umidade por até três meses. Usa-se para aplicar sobre feridas ou ainda, via oral, misturados ao leite, mel. Pode ser comido junto com as refeições.

Xarope

Cozinhar em um litro de água várias ervas indicadas para tosse e gripe. Deixe ferver bem. Caramelize duas xícaras de açúcar e junte ao cozimento. Espere esfriar e guarde em vidro esterilizado. Validade: 7 dias.
Tomar uma colher de sopa três vezes ao dia.

Eucalipto Tirodora: gripe, tosse (chá e banho). Usa-se no preparo de óleo (massagens e em gotas nos chás).
Gervão ou Folha Verde: infecção interna, intestino (chá). Obs: Não comer peixe.
Girassol: Febre de malária (chá das folhas secas), problemas pulmonares, doenças de estômago e resfriados.
Hortelã: vermes, queimaduras (óleo de oliva com hortelã), estomáquica.
Jasmim Borleta ou Jasmim Branco: pressão alta, garganta, dor no corpo, coração (água da flor), inflamação (raiz em garrafadas), depurativo.
Kiolô Cravo: coceira.
Linho: laxante, diurético.
Macela: febre e infecção intestinal. (chá e banho).
Malva: laxante, dor de dente, rins, bexiga, intestinos.
Malvarisco ou Altéia: laxante, calmante, expectorante (chá e xarope) e diurético (raiz macerada em água fria).
Manjericão: febre, inflamação.
Mastruz ou Erva Santa: feridas (aplicar folha machucada), vermes e tosse/tuberculose (tomar o sumo com leite); contusões (sumo com sal). As folhas verdes espalhadas pela casa espanta pulgas.
Melancia: infecção na urina (suco), febre (esmagar 9 sementes e verter água fervendo sobre elas, esfriar e tomar).
Melissa ou Erva-Cidreira de folha: calmante, antiespasmódico, distúrbio nervoso, nevralgia, hepática, fortificante do coração e cérebro. Associada ao Alecrim tem efeito antiinflamatório.
Mil Folhas: anti-hemorrágica, espasmos, adstringente, depurativa, hemorróidas.
Moscambé: catarro, gripe, pneumonia. Estomáquica, estimulante do aparelho digestivo e corrimento.
Nogueira: laxante, diurético (raiz). Leucorréia (folhas). Vermífugo (casca da noz).
Parietária: problema nas vias urinárias.
Pega Pinto: inflamação (raiz).
Romã: garganta (chá da casca do fruto), catarata (sumo que envolve as sementes).
Rosa Branca: depurativo, infecção no útero, calmante, laxante, bom para a vista.
Sabugueiro: sudorífero, laxante e diurético.
Salsa da Praia: depurativo, abcesso (aplicar folhas trituradas para supuração) e infecção.
Sálvia: cicatrizante, hipoglicemia, digestiva, diurética, anti-reumática, tônico do coração, adstringente, esgotamento nervoso, depressão (chá). Picada de inseto (esfregar as folhas verdes). Clarear os dentes e inflamação na gengiva (pilar folhas secas e usar para escovar os dentes).
Sucupira: anemia, febre reumática, problemas do coração, gripe, cansaço, bronquite e rouquidão. Usa-se as sementes. (quebrar 7 sementes, para garrafadas com biotônico. Pode renovar o biotônico até 3 vezes nas mesmas sementes). Tomar um cálice antes do almoço e

outro na janta durante 7 dias).
Tiririquinha: gripe, febre, pneumonia, tuberculose (torrar as sementes, triturar e peneirar até obter pó. Juntar ao pó de café, preparar normalmente e tomar).
Verônica: problemas digestivos, anticatarral. Suas sumidades floridas devem ser colhidas no Verão e secas ao Sol.
Violeta: catarros de bronquite, inflamação nas vias respiratórias.
Visco ou Parasita: antiespasmódica, moderadora da pressão arterial (folhas). Arteriosclerose, nefrite crônica, histeria e epilepsia (extrato).

PARTE III

RECEITAS

Xarope da Tidó

(tosse e rouquidão)

Raiz de tiririquinha, folhas de saião, gengibre, folhas de alfavaca e cravo.
Cortar as folhas, raspar e cortar as raízes. Colocar tudo para cozinhar bem, depois coar, colocar o açúcar meio queimado (caramelo), deixar esfriar e colocar em vidros esterilizados.

Pó para anemia

Usa-se também nas desnutrições.
Pega-se folhas de mandioca, lava-se e coloca-se para secar na sombra (meio sol) por 3 dias, ou na sombra por 5 dias.
Esquenta-se no forno por 5 minutos e soca-se no pilão até virar pó. Passar numa peneira bem fina.
Validade: 3 meses.
Dosagem: Crianças, 1 pitada no almoço e outra na janta
Adultos, 1 colher de sobremesa no almoço e na janta.
Pode-se usar também outras folhas como o Bredo.

Pó Antibiótico

Para infecções ou sobre feridas
Usa-se o mesmo processo do outro pó, só que com folhas de Maria Preta e flores da mesma.

Pomada de Arnica

Ferida, contusão, torção, dor de coluna
6 copos de óleo de arnica, preparado com antecedência, de preferência com óleo de coco. Derreta 2 copos de cera de abelha em panela de barro, mexendo com uma colher de pau. Retire do fogo e acrescente o óleo. Guardar em potinhos antes que endureça.
Validade: 2 anos

Garrafada de dona Dereta

Infecções em geral
Coloque um litro de água para ferver com um punhado de raiz de Purga da Praia (limpa e picada), raiz de um Jasmim Branco cortada e Canela em pau. Deixe ferver e coloque açúcar e ferva por mais alguns minutos. Deixe esfriar, coloque em um litro branco, sem coar. Enterre por 3 dias. Retire do chão e tome tudo durante 2 dias.

Vermífugo

Verminoses em geral
Sementes de abóbora, mamão, melão e melancia. Erva de Santa Maria (Mastruz) e amendoim.
Lavar as sementes e secar ao Sol. Juntar os demais itens, torrar no forno e triturar no liquidificador até obter uma paçoca.
Tomar o pó em jejum durante 3 a 4 dias e só se alimentar 20 minutos depois.
Dosagem: até 5 anos, 1 colher de chá
até 10 anos, 1 colher de sobremesa
adultos, 1 colher de sopa.
Repetir a dosagem após 20 dias.

Garrafada da dona Dereta

Calmante da tosse e para manipulação de pomadas.
Hidratante para pele e cabelo.
Para fazer 1 litro de óleo, usamos 10 cocos secos. Rale os cocos, passe água fervendo e esprema para retirar todo o leite. Repita esta operação 3 vezes. Deixe esfriar e coloque na geladeira até o dia seguinte. Retire com cuidado a banha que ficou por cima da vasilha e ponha numa panela. Deixe ferver bastante até apurar bem o óleo. Separe-o da borra que vai se formar no fundo da panela. Coe e deixe esfriar. Guarde em vidros esterilizados.
Se quiser um óleo mais claro, retire a pele do coco. O rendimento é menor.

PARTE IV

BANHOS MDICINAIS

São usados como tratamento complementar, principalmente pela sensação de bem estar que eles provocam. Podem ser jogados pelo corpo ou em bacias (para as crianças), de assento ou somente para os pés.

Gripe/Febre
Canema, Carqueja e Aroeira (o banho é quente, manter resguardo até o dia seguinte). Taririquinha, Cordão-de-Frade, Alfazema, Alfavaca, Eucalipto, Capim-Santo, Manjericão, Melão São Caetano. Seja qual for a erva escolhida, a água deve estar apenas dois graus a menos que a temperatura do corpo.

Alergias da pele
folha de Maracujá, folhas de Melancia, folhas de Sabugueiro.

Calmante
Alecrim, folha de Maracujá, Erva-Cidreira, Manjericão, Macela, Alfazema.

Problemas Genitais ((coceiras, inflamações e feridas)
Folha de Castanheira, casca de Cajueiro (o vermelho é melhor), Arrozinho, Rosa Branca (assento), Romã (para ferida no útero, fazer ducha interna), folhas de Framboesa e Erva Moura.

Malária
Folhas secas de Girassol.

Feridas
Alecrim, Confrei, Murta, Aroeira, Sálvia, Amor-Perfeito (cicatrizante), Eucalipto.

Icterícia
Picão, folhas de Lima (dar uns golinhos para a criança beber).

Estimulantes
Noz-Moscada, Cravo, Canela, Coentro e Angélica.

Hemorróidas
Casca de Cajueiro (vermelho), Mil-em-Ramos, casca de Carvalho, Cordão-de-Frade.

PARTE V

A CIA. SOMOS NÓS

Maria José Leandro de Meneses
Kátia Maria Neves Marques
Angela T. Noma
Loélio Ladeira
Maria José Maia (Zezé)
Maria de Lourdes Santana de Jesus
Dorotéia Batista Souto (Dona Dorota)
Magnólia Oliveira
Auzerina Batista (Zirinha)
Claudentina Trindade Alves (Dentina)
Adezina Maia dos Santos (Dezina)
Rita Lage (Ritinha)
Tidú
Jeferson
Plínio Mendonça
Conceicion Dias Costa
Ana Rita Souto dos Santos
Perminia Maia (Pepe)
Bernadete Maia Saproni (Bete)
Patrícia Uzelin (Pati)
Josefa Leandro de Meneses (Jô)
Simone de Lima Machado
Dona Aninha

AGRADECIMENTOS

As pessoas que devem ser agradecidas são as que trouxeram muito saber, as que sempre estiveram, as que se dedicaram de algum modo para que todos chegassem até aqui.
A Cia. Ofícios da Terra agradece a algumas pessoas e entidades que, com carinho e boa vontade, nos ajudaram bastante a realizar a melhor parte do nosso trabalho. Este primeiro volume das nossas conversas traz em si o agradecimento, inclusive, pelo o que virá.
Dona Tidú, dona Dorota, Dona Aninha, Zirinha, Dentina e Ritinha;
Parque Estadual de Itaúnas
Sociedade dos Amigos do Parque de Itaúnas (SAPI)

Obrigada

Printed by Books on Demand GmbH, Norderstedt / Germany